Sustainable Development Goals Series

The **Sustainable Development Goals Series** is Springer Nature's inaugural cross-imprint book series that addresses and supports the United Nations' seventeen Sustainable Development Goals. The series fosters comprehensive research focused on these global targets and endeavours to address some of society's greatest grand challenges. The SDGs are inherently multidisciplinary, and they bring people working across different fields together and working towards a common goal. In this spirit, the Sustainable Development Goals series is the first at Springer Nature to publish books under both the Springer and Palgrave Macmillan imprints, bringing the strengths of our imprints together.

The Sustainable Development Goals Series is organized into eighteen subseries: one subseries based around each of the seventeen respective Sustainable Development Goals, and an eighteenth subseries, "Connecting the Goals", which serves as a home for volumes addressing multiple goals or studying the SDGs as a whole. Each subseries is guided by an expert Subseries Advisor with years or decades of experience studying and addressing core components of their respective Goal.

The SDG Series has a remit as broad as the SDGs themselves, and contributions are welcome from scientists, academics, policymakers, and researchers working in fields related to any of the seventeen goals. If you are interested in contributing a monograph or curated volume to the series, please contact the Publishers: Zachary Romano [Springer; zachary.romano@springer.com] and Rachael Ballard [Palgrave Macmillan; rachael.ballard@palgrave.com].

Muhammad Mushahid Anwar
Asad Aziz
Editors

Sustainable Landscape for Urban Resilience

Achieving the UN Sustainable Development Goals

Springer

Editors
Muhammad Mushahid Anwar
Department of Geography
University of Gujrat
Gujrat, Pakistan

Asad Aziz
Department of Geography
University of Gujrat
Gujrat, Pakistan

ISSN 2523-3084 ISSN 2523-3092 (electronic)
Sustainable Development Goals Series
ISBN 978-3-031-99835-5 ISBN 978-3-031-99836-2 (eBook)
https://doi.org/10.1007/978-3-031-99836-2

This Springer imprint is published by the registered company Springer Nature Switzerland AG
The registered company address is: Gewerbestrasse 11, 6330 Cham, Switzerland

Competing Interests The authors have no competing interests to declare that are relevant to the content of this manuscript.

Contents

Introduction: Framing Sustainable Landscape for Urban Resilience

Muhammad Mushahid Anwar

Abstract

The twenty-first century marks an era of accelerated urbanization, placing cities at the epicenter of both innovation and crisis. Urban areas are now home to more than half of the global population, and this number is projected to rise to nearly 68% by 2050 (United Nations, World urbanization prospects: The 2018 revision. UN Department of Economic and Social Affairs, 2019). While this trend signifies progress in terms of economic activity, access to services, and modernization, it has also generated unprecedented environmental and social challenges. The physical expansion of cities, often driven by informal settlements and speculative development, has resulted in land degradation, biodiversity loss, increased energy demands, and declining quality of life.

This expansion is often chaotic and unsustainable, especially in developing countries where urban planning frameworks either do not exist or are poorly enforced. Urban sprawl consumes agricultural land, depletes green areas, and intensifies the demand for infrastructure and services that city governments struggle to provide. In many rapidly urbanizing contexts, environmental degradation, rising urban heat islands, flooding, traffic congestion, and poor waste management have become defining features of the urban experience.

At the same time, the climate crisis adds another layer of vulnerability. Cities are disproportionately affected by climate-induced hazards such as extreme heat, floods, droughts, and rising sea levels. As urban populations grow, so does their exposure to these risks. In South Asia, for instance, cities like Karachi, Lahore, and Islamabad are experiencing longer, hotter summers and unpredictable monsoonal flooding, while cities in Eastern Europe like Bucharest face seasonal droughts and heat stress due to reduced green infrastructure and dense construction.

The traditional approaches to urban planning, which prioritized car-centric infrastructure and economic growth over environmental and social concerns, are no longer viable. These outdated models have failed to account for the complex interdependencies between urban systems and the ecosystems they inhabit. Moreover, they often perpetuate socio-economic inequalities by ignoring the needs of marginalized communities, who are typically the most vulnerable to climate impacts and environmental degradation.

In response, the concept of urban resilience has gained momentum in policy, planning,

M. M. Anwar (✉)
Department of Geography, University of Gujrat, Gujrat, Punjab, Pakistan
e-mail: mushahid.anwar@uog.edu.pk

M. M. Anwar, A. Aziz (eds.), *Sustainable Landscape for Urban Resilience*, Sustainable Development Goals Series, https://doi.org/10.1007/978-3-031-99836-2_1

and academic discourse. Urban resilience refers to the capacity of a city to absorb, recover from, and adapt to a range of shocks and stresses, whether environmental, economic, or social (Meerow et al., Landsc Urban Plan 147:38–49, 2016). However, resilience is not just about technical solutions or infrastructure upgrades. It encompasses a broader transformation in how cities are conceptualized and governed—emphasizing flexibility, redundancy, diversity, and inclusivity.

This chapter, and the book as a whole, argues that achieving urban resilience requires more than risk mitigation. It calls for a fundamental shift in how we design, build, and inhabit urban spaces. This includes rethinking the role of natural systems within cities, integrating ecological principles into planning, and ensuring that the benefits of sustainability are equitably distributed. It is within this context that sustainable landscape planning emerges as a critical tool—linking environmental, social, and spatial dimensions of resilience in ways that are adaptive, inclusive, and forward-looking.

Keywords

Adaptive · Landscapes · Sustainable · Sustainable landscape · Resilience

Sustainable Landscapes: Definitions and Dimensions

The term sustainable landscape encompasses more than just green parks and recreational spaces—it refers to the integrated design, planning, and management of urban and peri-urban spaces to meet the environmental, social, and economic needs of present and future generations. Sustainable landscapes are multifunctional systems that go beyond aesthetics to support ecosystem services, promote biodiversity, regulate microclimates, manage water, and enhance the well-being of communities. As cities grapple with rapid population growth, climate change, and socio-spatial inequities, the role of landscapes in fostering urban resilience becomes increasingly indispensable.

The conceptual foundation of sustainable landscapes draws from multiple disciplines, including landscape ecology, urban planning, sustainability science, and environmental design. Landscape ecology emphasizes the spatial arrangement of ecosystems and their functioning across scales, offering insights into how patches of vegetation, water bodies, and built environments interact. Urban planning brings in spatial logic and zoning systems to organize land use, while sustainability science urges the integration of long-term ecological balance with socio-economic development goals. Together, these fields inform a holistic approach to urban development that values both ecological integrity and human livability.

Sustainable landscapes provide provisioning, regulating, supporting, and cultural ecosystem services. For instance, green infrastructure such as urban forests and wetlands helps regulate air temperature, reduce runoff, improve air and water quality, and sequester carbon. These regulating services are especially critical in mitigating the urban heat island (UHI) effect a phenomenon where dense urban areas become significantly hotter than surrounding rural regions due to heat-retaining materials and lack of vegetation. Provisioning services include food and medicinal plant sources, especially through urban agriculture. Supporting services like pollination and soil formation help sustain biodiversity and ecological functioning. Lastly, cultural services—such as recreational opportunities, aesthetic experiences, and spiritual connection to nature—enhance the mental and emotional health of urban residents.

Importantly, sustainable landscapes also offer a social function. Well-designed public green spaces have been linked to higher levels of social cohesion, lower crime rates, and improved physical and mental health outcomes. In dense and economically stratified cities, they act as "social equalizers," providing free and accessible spaces for diverse communities to interact. Inclusive landscape planning, therefore, becomes a means to address social injustices, especially in areas where marginalized groups lack access to clean, safe, and well-maintained green spaces.

From a planning perspective, landscape morphology—the spatial configuration and interconnectivity of green and blue infrastructure—is critical. Cities with well-connected green corridors, riparian buffers, and networks of parks demonstrate better environmental performance than those with isolated green patches. Islamabad's once-pristine natural zones have been fragmented due to real estate expansion, while Bucharest struggles with uneven green space distribution across its administrative sectors. These examples show that sustainability is not just a matter of quantity but also quality, accessibility, and connectivity of landscapes.

In sum, sustainable landscapes are living systems that must be strategically integrated into the urban fabric. They function not only as environmental regulators but also as social infrastructure, economic assets, and resilience mechanisms. Their design and governance must therefore be informed by scientific evidence, local knowledge, and inclusive policymaking a central theme explored throughout this volume.

The Role of Technology: GIS, NDVI, and UAV Integration

As urban landscapes become more complex and dynamic, the demand for robust analytical tools to understand and manage land use, environmental conditions, and spatial inequality has grown significantly. In this context, geographic information systems (GIS), normalized difference vegetation index (NDVI), and unmanned aerial vehicles (UAVs) have emerged as powerful technologies for data-driven urban landscape planning. These tools enable researchers, planners, and decision-makers to visualize, monitor, and predict spatial changes over time with precision—supporting evidence-based strategies for sustainability and resilience.

GIS is the foundational technology for urban spatial analysis. It allows users to collect, store, manage, and analyze layers of spatial and attribute data, which can include land use classifications, topography, zoning regulations, infrastructure networks, and demographic information. GIS plays a central role in land use planning, environmental modeling, risk assessment, and service delivery optimization. In urban sustainability studies, GIS is used to map green infrastructure, identify urban heat islands, model flood risks, and determine areas with poor accessibility to parks or public amenities.

In Islamabad, for example, GIS-based classification of satellite imagery from 2000 to 2020 has shown a rapid expansion of built-up areas at the expense of green and agricultural land (Aziz & Anwar, 2024). By analyzing temporal data layers, planners can visualize how unregulated housing schemes and peri-urban sprawl are altering the city's morphology. These insights also support policy enforcement by identifying illegal encroachments or violations of the city's Master Plan.

NDVI—a remote sensing index derived from satellite imagery—is a key tool in measuring vegetative health and density. The NDVI formula uses the differential reflectance of near-infrared and visible red light to assess whether a given area contains live green vegetation. Values range from -1 to $+1$, with higher values indicating healthier, denser vegetation. NDVI is crucial for assessing the ecological performance of urban areas and for identifying hotspots of degradation or deforestation.

In urban studies, NDVI is often overlaid on administrative boundaries or land use maps to track vegetation trends across different neighborhoods. This technique has revealed, for example, that in both Islamabad and Bucharest, vegetative cover is significantly lower in low-income or informal settlements. Such findings highlight environmental inequality and provide a scientific basis for targeting green infrastructure investments.

UAVs—commonly referred to as drones—have revolutionized data collection by providing high-resolution aerial imagery and real-time environmental monitoring. Unlike satellites, UAVs can fly at lower altitudes, capture images with higher spatial detail, and be deployed rapidly in localized contexts. In Bucharest, UAVs equipped with thermal cameras were used to assess land surface temperature (LST) differences across neighborhoods with varying levels of green coverage (Gavrilidis et al., 2022). The results confirmed that Small Public Urban Green Spaces (SPUGS) significantly reduce surface temperatures, thereby offering localized climate regulation benefits.

UAVs are also useful for 3D modeling, topographic mapping, and even assessing vegetation health through multispectral imaging. These features make drones especially valuable in regions where satellite data is limited, outdated, or expensive. They also support participatory mapping exercises, where local communities can be involved in identifying assets and vulnerabilities in their neighborhoods.

Together, GIS, NDVI, and UAV technologies form a triad of spatial intelligence that enhances our ability to understand urban landscapes not just in physical terms, but also in relation to social, ecological, and climatic processes. Their integration into planning workflows represents a paradigm shift—from reactive governance based on assumptions to proactive, data-informed decision-making. This technological toolkit, explored in-depth in later chapters, is essential for designing sustainable urban landscapes that are resilient, equitable, and adaptive to future uncertainties.

Case Study: Islamabad—Urban Expansion and Ecological Fragmentation

Islamabad, the capital city of Pakistan, was originally designed in the 1960s as a model of planned urban development, inspired by modernist principles that prioritized orderly zoning, ecological preservation, and a strong relationship between built and natural environments. The city was divided into distinct sectors, each with designated functions—residential, commercial, diplomatic, institutional—and buffered by green spaces and natural landscapes such as the Margalla Hills and the Rawal Lake. For decades, Islamabad was considered one of South Asia's most livable cities due to its low population density, clean air, and proximity to nature. However, in recent years, the city has undergone a dramatic transformation, marked by unregulated urban expansion and the encroachment of ecological buffers.

Using multitemporal GIS-based land use analysis, researchers have documented significant changes in Islamabad's land cover between 2000 and 2020. Built-up areas, especially in Zones IV and V, have increased substantially, while green cover and agricultural land have diminished at an alarming rate (Aziz & Anwar, 2024). The expansion is most pronounced along the city's peripheries, where housing societies and informal settlements have proliferated, often outside the parameters of the city's Revised Master Plan (2020–2040). These developments have fragmented the once-continuous ecological corridors that linked Islamabad to the surrounding natural environment, reducing the city's overall landscape connectivity and resilience.

One of the main drivers of this expansion is the real estate boom, fueled by rising demand for housing, speculative investment, and weak planning enforcement. Several private housing societies have been developed on ecologically sensitive land, including forested areas and catchment zones. While some of these developments are legal, many are not approved by the Capital Development Authority (CDA), yet continue to operate due to loopholes in governance and political patronage. As a result, the urban fabric has expanded in a fragmented, uncoordinated manner that undermines ecological sustainability.

The environmental consequences are severe (Meerow et al., 2016). The removal of vegetative cover has intensified the urban heat island (UHI) effect, increased surface runoff, and raised the risk of flash floods—particularly during the monsoon season. Biodiversity loss is another concern, as many wildlife corridors have been severed by new roads and construction. Air quality has deteriorated due to increased vehicular traffic and the reduction of tree cover, while groundwater levels continue to drop amid unregulated extraction and inadequate recharge mechanisms.

Moreover, the socio-spatial implications of Islamabad's growth are significant. Informal settlements, often lacking access to clean water, sanitation, and waste management, are expanding rapidly in peri-urban zones (United Nations, 2019). These communities are highly vulnerable to environmental hazards and are often excluded from planning processes. At the same time, high-income gated societies with manicured landscapes create a sharp contrast—reinforcing spatial segregation and environmental injustice.

In response to these issues, the city's planners have proposed revisions to the Islamabad Master Plan, emphasizing sustainable urban development, ecological restoration, and stricter zoning compliance. However, implementation remains weak. Enforcement mechanisms, institutional coordination, and political will are often lacking, and environmental considerations are frequently subordinated to short-term economic interests.

The case of Islamabad illustrates a broader global trend: how cities planned for sustainability can quickly become unsustainable when market forces, population pressures, and governance failures collide. It also highlights the critical role of landscape planning and regulation in safeguarding urban resilience—an issue that this book explores in further depth through comparative analysis.

Case Study: Bucharest—Green Space Inequality and Microclimatic Benefits

Bucharest, the capital city of Romania, presents a unique case of urban transformation shaped by its socialist legacy and post-socialist restructuring. As the largest city in Romania and one of the most densely populated urban centers in Eastern Europe, Bucharest has undergone significant changes since the fall of communism in 1989. While the post-socialist era has brought economic growth, infrastructure upgrades, and new investment, it has also resulted in spatial fragmentation, green space loss, and growing socio-environmental inequalities. These changes have critical implications for the sustainability and resilience of the city's landscape.

Under the centralized planning model of the communist era, Bucharest's urban structure was characterized by dense residential blocks, centralized services, and a few large parks and public green areas. However, during the transition to a market-based economy, public assets—including green spaces—were often privatized or poorly managed. Informal construction, speculative real estate development, and a lack of cohesive urban policy led to a disorganized and uneven urban expansion. Today, the city suffers from an unbalanced distribution of green spaces, with some districts enjoying tree-lined boulevards and accessible parks, while others, particularly working-class or marginalized neighborhoods, remain severely underserved.

Quantitative assessments using UAV-based thermal sensors and GIS mapping have revealed stark microclimatic differences across the city. Areas with more vegetation and better-maintained public parks tend to have significantly lower land surface temperatures (LST)—by 2–3 °C during summer peaks compared to highly built-up and vegetation-deficient districts (Gavrilidis et al., 2022). This difference is not just a matter of comfort; it directly affects residents' health, particularly for vulnerable groups such as the elderly, children, and those with preexisting health conditions.

One of the most insightful findings from recent studies is the role of Small Public Urban Green Spaces (SPUGS)—pocket parks, neighborhood gardens, and small wooded patches. Though often less than 0.5 hectares in size, these green areas have shown measurable impacts in reducing local temperatures, improving air quality, and providing accessible social spaces. SPUGS are particularly vital in densely populated urban environments where large green parks are either distant or overburdened.

However, access to these green resources is far from equitable. Residents in wealthier sectors such as Sector 1 or parts of Sector 2 benefit from tree-lined streets and park proximity, while lower-income districts like Sector 5 or Ferentari struggle with less than 3 m^2 of green space per capita—far below the European Environment Agency's recommended minimum of 9 m^2. This spatial disparity reflects deeper issues of environmental injustice, where socio-economically disadvantaged populations face a disproportionate share of environmental burdens and are excluded from the benefits of urban greening.

Municipal efforts have been made to reverse these trends, including tree planting programs, rehabilitation of abandoned lots into parks, and partnerships with community groups. However, the success of these initiatives remains mixed due to fragmented governance structures, lack of sustained funding, and minimal citizen participation in decision-making. Moreover, conflicts often arise between development priorities and environmental preservation, especially in areas where land values are rising.

The Bucharest case demonstrates how micro-level interventions in urban green space planning can yield significant macro-level benefits—especially when guided by spatial analysis and community needs. It also reveals that green infrastructure planning must account for social equity, ensuring that all residents—regardless of income, ethnicity, or neighborhood—have access to the ecological, climatic, and social benefits that green spaces provide.

Socio-Spatial Disparities and Equity in Urban Landscapes

Urban landscapes are not just physical spaces—they are socio-political constructs that reflect and reinforce the underlying power dynamics of a city. How green spaces, infrastructure, and environmental resources are distributed across an urban area often mirrors broader patterns of social inclusion or exclusion. The spatial arrangement of green infrastructure can significantly impact public health, environmental quality, mobility, and social cohesion. Yet, many cities around the world—including Islamabad and Bucharest—exhibit deep socio-spatial disparities that prevent marginalized communities from accessing the benefits of sustainable landscapes.

Socio-spatial disparities refer to the unequal distribution of physical and social amenities across different geographic areas, often linked to class, ethnicity, or historical marginalization. In many rapidly growing cities, affluent neighborhoods enjoy tree-lined boulevards, gated parks, and well-maintained public facilities, while poorer or informal areas suffer from degraded environments, pollution, and inadequate services. These inequalities are not accidental; they are the outcomes of policy decisions, planning frameworks, and market forces that systematically prioritize some areas over others.

In Islamabad, as detailed earlier, unregulated real estate expansion has disproportionately benefited high-income groups. Housing schemes with private landscaping, water supply, and waste services proliferate along the city's periphery, while informal settlements on the outskirts lack even basic sanitation. The shrinking of public green areas—once considered common goods—is particularly harmful to low-income families who cannot afford private recreational or cooling spaces. Additionally, new infrastructure often bypasses these communities, deepening their physical and social isolation.

Similarly, in Bucharest, historical zoning decisions, post-socialist fragmentation, and weak municipal coordination have created spatial inequalities in access to urban nature. Districts like Ferentari have minimal green space coverage, poor maintenance of public infrastructure, and limited public investment. In contrast, central and affluent districts benefit from large parks, clean streets, and heritage preservation. These disparities are further compounded by differences in political representation, civic engagement, and institutional capacity across districts.

Such unequal spatial configurations can lead to environmental injustice, where disadvantaged groups bear a disproportionate share of environmental burdens (e.g., heat stress, air pollution) while enjoying fewer benefits (e.g., tree canopy, access to recreational areas). Environmental justice advocates argue that planning systems must go beyond physical metrics and include procedural justice (equal participation in decision-making), distributional justice (fair resource allocation), and recognitional justice (acknowledging the rights and voices of marginalized groups) (Agyeman et al., 2003).

Addressing these disparities requires more than technical interventions—it calls for a reorientation of urban planning toward equity and inclusion. This can involve participatory GIS mapping with local communities to identify needs and priorities, developing urban greening programs in under-resourced areas, and enforcing legal frameworks that guarantee equitable distribution of environmental resources. Moreover, policymakers must consider intersectional factors such as gender, age, and disability, which influence how people use and perceive public landscapes.

An equitable urban landscape is not only a moral imperative but also a resilience strategy. Research shows that socially cohesive and inclusive neighborhoods recover faster from crises, adapt better to climate change, and experience lower rates of violence and disease. Integrating equity into sustainable landscape planning ensures that urban resilience is truly collective—and that no community is left behind.

Nature-Based Solutions (NBS) for Urban Adaptation

In the face of intensifying urbanization and climate change, cities are increasingly turning to nature-based solutions (NBS) as a sustainable, cost-effective, and multifunctional approach to address urban environmental challenges. NBS are defined by the European Commission as "solutions that are inspired and supported by nature, which are cost-effective, simultaneously provide environmental, social and economic benefits, and help build resilience." Rather than relying solely on gray infrastructure—such as concrete drainage systems, air conditioning, and road expansion—NBS harness the power of natural systems to solve urban problems while enhancing ecological health and human well-being.

NBS include a wide range of interventions, from urban forests, wetlands, and river restoration to green roofs, permeable pavements, and bioswales. These interventions offer tangible benefits such as cooling urban microclimates, filtering pollutants, managing stormwater, reducing noise, and supporting biodiversity. In cities like Islamabad, where monsoon-related flooding and rising urban heat are growing concerns, and Bucharest, where microclimatic variation and green space inequality persist, NBS can serve as critical tools for climate adaptation and spatial equity.

Urban forests and tree-lined streets, for example, help mitigate the urban heat island (UHI) effect by providing shade and facilitating evapotranspiration, which cools the surrounding air. Green roofs and walls improve building insulation, reduce energy consumption, and enhance urban aesthetics. Rain gardens and bioswales—vegetated channels designed to absorb runoff—prevent urban flooding and recharge groundwater. When integrated into urban planning, these systems reduce the burden on traditional drainage and cooling infrastructure, offering cities a more resilient and flexible approach to climate hazards.

Importantly, NBS are not only ecologically beneficial—they are socially transformative. They enhance the livability of cities, support public health, and promote environmental education. Green spaces have been shown to reduce stress, encourage physical activity, and improve mental health. Community gardens, green alleys, and pocket parks foster social interaction and intergenerational bonding, particularly in dense urban neighborhoods where space is limited.

In Islamabad, certain community-led greening projects—such as green belts along major roads and tree plantation drives—have shown the potential for scalable NBS initiatives. However, many of these efforts remain uncoordinated and lack long-term institutional support. Similarly, in Bucharest, there is growing interest in integrating green infrastructure with public transportation corridors and vacant lots, yet implementation is slow due to bureaucratic fragmentation and competing urban development priorities.

One of the major challenges to mainstreaming NBS is the lack of interdisciplinary collaboration among planners, ecologists, engineers, economists, and community stakeholders. Additionally, decision-makers often underestimate the economic value of ecosystem services or view green interventions as supplementary rather than essential. There is also a need for clear policy frameworks, funding mechanisms, and monitoring tools to evaluate the performance and maintenance of NBS over time.

Globally, the discourse on NBS is gaining traction through initiatives like the UN Decade on Ecosystem Restoration (2021–2030) and the European Green Deal, which encourage cities to adopt nature-based strategies. These global trends must be localized and contextualized to reflect the specific socio-political, climatic, and spatial realities of each city. For Islamabad and Bucharest, this means recognizing NBS not only as environmental interventions but as catalysts for inclusive urban transformation.

In conclusion, nature-based solutions offer an integrative pathway to build climate resilience, restore ecological balance, and improve urban quality of life. When strategically planned, equitably distributed, and effectively governed, NBS can reshape urban futures in ways that are sustainable, just, and regenerative.

A Framework for Resilience Through Sustainable Landscape Planning

The challenges faced by modern cities—climate change, ecological degradation, socio-economic inequality, and infrastructural stress—require more than isolated solutions. Instead, what is needed is an integrated, systems-based framework that places sustainable landscapes at the heart of urban resilience. Such a framework must balance ecological functionality, spatial planning, community participation, and technological innovation to prepare cities for long-term sustainability and adaptation.

This section proposes a conceptual and practical framework structured around four interconnected pillars: (1) ecological integrity, (2) technological integration, (3) institutional governance, and (4) community engagement. Each of these pillars addresses a specific set of urban vulnerabilities and contributes to building a holistic response to socio-environmental challenges.

Ecological Integrity

The foundation of any sustainable urban system is its ecological health. This includes the conservation and restoration of biodiversity, natural habitats, and ecosystem services. Urban areas must be viewed as part of larger ecological networks, where green corridors, wetlands, forests, and open spaces function as critical infrastructure. In Islamabad, the fragmentation of vegetative buffers and unchecked development in conservation zones illustrate the consequences of ignoring ecological thresholds. In contrast, the protection and expansion of Bucharest's green nodes, such as SPUGS and urban parks, reveal how even small-scale ecological interventions can improve urban resilience.

Technological Integration

The second pillar involves the strategic use of spatial and environmental technologies such as GIS, NDVI, UAVs, IoT-based sensors, and remote sensing platforms. These tools provide real-time, fine-scale data that can be used to map land use change, analyze thermal stress, track vegetation health, and model urban growth scenarios. Importantly, technology enables cities to shift from reactive to proactive planning. Predictive modeling, for instance, can forecast flood-prone zones or overheating risks under various climate change trajectories, allowing for timely mitigation strategies. However, for this pillar to succeed, it must be paired with trained personnel, accessible data systems, and a clear mandate for evidence-based decision-making.

Institutional Governance

Policy coherence, regulatory enforcement, and administrative capacity form the third pillar of resilience. The absence of effective governance often leads to plan violations, environmental degradation, and social exclusion. In Islamabad, multiple planning authorities with overlapping mandates lead to confusion and loopholes exploited by private developers. In Bucharest, fragmented institutional responsibilities across districts result in inconsistent green space standards. A resilience-focused governance structure must be adaptive, transparent, and cross-sectoral, capable of integrating environmental concerns into every aspect of urban development—from transportation and housing to health and education.

Community Engagement and Equity

Resilience cannot be imposed from the top down; it must be co-created with communities. This final pillar centers on participatory planning, community stewardship of green spaces, and recognition of marginalized voices in urban decision-making. Cities are most resilient when their residents are empowered to act—through environmental education, volunteer greening efforts, or citizen science. Equity is a critical concern here: resilience must benefit all, especially the most vulnerable. Mechanisms for inclusion must be embedded in planning processes, ensuring fair access to resources, services, and opportunities.

This four-pillar framework is not a fixed model but a guiding lens through which cities can diagnose challenges, design interventions, and measure progress. It aligns with global agendas like the New Urban Agenda, the Sendai Framework for Disaster Risk Reduction, and the Sustainable Development Goals (SDGs), particularly SDG 11 (Sustainable Cities), SDG 13 (Climate Action), and SDG 15 (Life on Land). Through the lens of Islamabad and Bucharest, this book demonstrates how such a framework can be tailored and applied in diverse geographic, political, and socio-economic contexts.

Structure of the Book

This book, *Sustainable Landscape for Urban Resilience*, is structured to guide readers through an interdisciplinary journey that begins with theory, moves into applied analysis, and culminates in cross-case synthesis and actionable recommendations. It is designed for a broad audience urban planners, researchers, policymakers, and graduate students interested in understanding how landscape thinking can address the complex challenges of sustainability and resilience in the contemporary urban context.

The book is divided into 10 chapters, each offering a unique perspective while maintaining thematic continuity across the case studies of Islamabad and Bucharest. These two cities serve as the empirical anchors for the book's arguments. Despite their contrasting geographies, political systems, and urban histories, both cities face strikingly similar issues ecological degradation, spatial inequality, institutional fragmentation, and mounting climate vulnerabilities. Their comparison helps identify universal lessons and context-specific strategies that can inform other cities facing similar challenges.

Chapter 1

This introductory chapter provides the theoretical and methodological foundations for the rest of the book. It defines key terms such as sustainable landscapes, resilience, socio-spatial equity, and nature-based solutions and introduces the analytical tools used throughout, including GIS, NDVI, and UAV-based analysis. It also presents the conceptual framework that guides the book's analysis: one based on ecological integrity, technology integration, governance reform, and participatory equity.

Chapters 2–4

These chapters are devoted to Islamabad, Pakistan's planned capital city, which now faces unregulated expansion and ecological fragmentation.

- Chapter 2 uses satellite data and GIS to map land use changes from 2000 to 2020, analyzing the drivers and implications of spatial transformation.

- Chapter 3 explores the socio-economic sustainability of different sectors of the city, highlighting disparities in service provision, environmental quality, and housing accessibility.
- Chapter 4 focuses on the role of green infrastructure, assessing its coverage, function, and vulnerability across administrative zones.

Together, these chapters show how a city originally designed for sustainability can drift toward unsustainable outcomes in the absence of effective governance and citizen engagement.

Chapters 5–7

These chapters turn to Bucharest, Romania's capital, as a case of post-socialist urban transition.

- Chapter 5 examines urban morphology and fragmentation, highlighting the impact of historical zoning and privatization on green space distribution.
- Chapter 6 presents data from UAV-based thermal and spatial analysis to demonstrate the cooling and social benefits of Small Public Urban Green Spaces (SPUGS).
- Chapter 7 investigates spatial equity, revealing significant disparities in green access and participatory governance across different city sectors.

The Bucharest case provides insight into the legacy effects of centralized planning and the challenges of reforming urban systems in the face of political and economic decentralization.

Chapters 8 and 9

These thematic chapters synthesize findings from both case studies, drawing parallels and contrasts. Chapter 8 focuses on governance structures, policy gaps, and the role of citizen initiatives. Chapter 9 outlines a comparative landscape planning model that includes resilience metrics, monitoring indicators, and scaling strategies adaptable to different global regions.

Chapter 10

This chapter is related to the role of blue infrastructure, that is, the water bodies in urban areas. The water bodies are a component of ecosystem that play a vital role in the sustainability of environment and balancing of ecosystem services. This chapter provides a snapshot of this concept.

Chapter 11

This chapter is related to the pollution of heavy metals, particularly Pb and Zn, taking China as an example. The industrialization in the region is gradually increasing that also negatively impacts the local environment and increasing the air pollution, including the heavy metals.

Chapter 12

This chapter provides a thematic mapping of citizen attitude for environmental, economic, and social services in urban areas. Due to the increasing rate of urbanization, the pressure on these services has increased. The regions in mega cities suffer due to the lack of provision of these services. This chapter provides a general perception of citizens regarding the environmental, social, and economic opportunities in the cities.

Chapters 13 and 14

Both of these chapters are associated with the impacts of climate change, resilience, and current progress in mitigations around the world. Chapter 13 discusses the case of production of wheat and rice as agricultural production and impacts of climate change on agriculture in the region, that is, Pakistan. While Chap. 14 elaborates the urban resilience and climate change current goals and future adoption and mitigations strategies to tackle the impacts of climate change for the future.

Chapter 15

Chapter 15 presents an integrative conclusion and policy roadmap. It reflects on the conceptual and empirical lessons of the book and offers recommendations for future research, practice, and institutional reform anchored in APA-style references and global policy frameworks.

Conclusion

The transformation of cities into resilient, inclusive, and ecologically sustainable systems is one of the most pressing challenges of the twenty-first century. As urban populations grow and climate risks intensify, the way we plan, manage, and inhabit urban spaces must evolve to address a wide range of socio-environmental vulnerabilities. This chapter has introduced the core concepts, frameworks, tools, and case studies that shape this book's approach to sustainable landscape planning for urban resilience.

The chapter begins by outlining the global context of urbanization and sustainability crises, where unchecked expansion, ecosystem degradation, and rising inequality have placed enormous stress on both natural systems and institutional capacities. In response, the concept of urban resilience was introduced not merely as a strategy for survival, but as a transformative paradigm for long-term sustainability, adaptability, and justice.

Central to this transformation is the idea of sustainable landscapes—integrated systems that provide ecological services, cultural value, and socio-spatial equity. The chapter detailed how landscapes can serve as both infrastructure and interface: infrastructure for delivering climate mitigation and ecosystem benefits, and interface between people and their environments, fostering public health, civic identity, and inclusive governance.

Through the lens of technological tools like GIS, NDVI, and UAVs, the book embraces an evidence-based planning approach. These tools provide spatial intelligence that allows planners to analyze past trends, visualize present patterns, and simulate future scenarios. They also support policy accountability and public engagement by making environmental data more accessible and actionable.

The case studies of Islamabad and Bucharest exemplify how differing political and planning contexts produce unique but convergent challenges. Islamabad, despite its origins as a planned city, suffers from informal development, poor enforcement, and ecological fragmentation. Bucharest, shaped by post-socialist transition, faces institutional fragmentation and green space inequity. Yet both cities demonstrate how targeted interventions—such as nature-based solutions, participatory mapping, and governance reforms—can lead to more resilient outcomes.

This chapter also introduces a four-pillar framework—ecological integrity, technological integration, institutional governance, and community engagement—as a foundation for planning sustainable landscapes. It highlights that urban resilience is not only a technical exercise, but a political and ethical commitment to fair distribution of resources, meaningful public participation, and long-term environmental stewardship.

Finally, the chapter outlines the structure of the book, guiding readers from foundational theory to applied research and comparative insights. By combining spatial analysis, empirical case work, and policy discourse, the book aims to make a meaningful contribution to the emerging field of landscape-based urban resilience planning.

As cities continue to grow in size, complexity, and vulnerability, sustainable landscapes offer a tangible and hopeful pathway toward regenerative futures. But realizing that potential requires visionary leadership, cross-sector collaboration, and a strong scientific and ethical grounding—themes that are explored in greater detail in the chapters to follow.

References

Agyeman, J., Bullard, R. D., & Evans, B. (2003). *Just sustainabilities: Development in an unequal world*. MIT Press.

Aziz, A., & Anwar, M. M. (2024). Assessing the impacts of urban expansion on sustainability; empirical evidence from Islamabad, Pakistan. *GeoJournal, 89*(2), 82. https://doi.org/10.1007/s10708-024-11087-6

Gavrilidis, A.-A., Popa, A.-M., Onose, D. A., & Grădinaru, S. R. (2022). Planning small for winning big: Small urban green space distribution patterns in an expanding city. *Urban Forestry & Urban Greening, 78*, 127787.

Meerow, S., Newell, J. P., & Stults, M. (2016). Defining urban resilience: A review. *Landscape and Urban Planning, 147*, 38–49.

United Nations. (2019). *World urbanization prospects: The 2018 revision*. UN Department of Economic and Social Affairs.

Mapping Land Use Dynamics and Their Impacts on Sustainability: A Study of Islamabad Pakistan

Asad Aziz and Muhammad Mushahid Anwar

Abstract

Sustainable land use in an urban area is the fundamental need of the day for planned urbanization. To estimate urban land use, GIS uses the past two decades' data to map the total change in land use patterns, changes in land surface temperature, and changes in vegetation cover in the city. Meanwhile, multilinear regression analysis was used for economic and social sustainability measures. For environmental sustainability, the study used land use classification and GIS analysis, including measurement of land surface temperature (LST) and Normalized Difference Vegetation Index (NDVI), while the revised master plan, urban newspaper, and public opinion were used as secondary data to estimate socioeconomic sustainability measures. The regression values for socioeconomic sustainability measures through the coefficient of regressions $R = 0.745$ and $R = 0.991$ show the alarming situation of sustainability and its challenges. The research concludes that there is a lack of planning in urban land use and city expansion, which poses major threats to urban development and sustainability in the region. It is recommended that the relevant authorities, for example, Capital Development Authority, should focus on the implementation of a revised master plan to regulate unplanned urban expansion for city sustainability.

Keywords

City · Development · Expansion · GIS · Islamabad · Land use · Urban planning · United Nations · UN · Sustainability

Introduction

Land use planning of a city is a fundamental principle for sustainable spatial urban planning and urbanization. Sustainable land use planning is imperative for growing populated regions, as in the case of South Asia or globally (Arfanuzzaman & Dahiya, 2019). A city with planned land use will lead to sustainable urban development that ensures regional development and a high quality of life as a result of maximizing environmental services (Kalfas et al., 2023). In developing nations, rapid population growth and migration from rural to urban settings put significant pressure on natural resources, including land and land cover, and the same situation is observed in the study area, Islamabad Pakistan. This unplanned and rapid urbanization disturbs the spatial pattern of urban land and challenges the spatial sustain-

A. Aziz (✉) · M. M. Anwar
Department of Geography, University of Gujrat, Gujrat, Punjab, Pakistan
e-mail: asadaziz@uog.edu.pk; mushahid.anwar@uog.edu.pk

M. M. Anwar, A. Aziz (eds.), *Sustainable Landscape for Urban Resilience*, Sustainable Development Goals Series, https://doi.org/10.1007/978-3-031-99836-2_2

ability of the area (Li et al., 2016; Zhou et al., 2022). For sustainable development it is essential to address this pattern of unplanned urban land use and formulate the planning with an eye toward regional development and spatial prosperity (Mallick et al., 2021). Additionally, land use planning and other spatial environmental parameters, for example, land surface temperature (LST), Normalized Difference Vegetation Index (NDVI), solid waste management, environmental justice, smart energy, and public transport, are also treated as current needs in achieving UN Sustainability Development Goals in urban settings (Mahadevia et al., 2023). Previous studies showed that another major cause of unplanned urban expansion was the housing scheme introduced by the land mafia not only in Pakistan but even in the Global South and South Asia during the twenty-first century (Benjamin, 2015; Pandey & Asif, 2022).

This study was designed to estimate the total land use change in the past two decades in the capital city of Pakistan. Meanwhile, socioeconomic challenges faced by the local population were taken to be economic and social sustainability. In the city, the growing population, illegal property brokers, and poor land mangement are considered to be the major causes of environmental degradation threatening urban sustainability in the region (Roy, 2012).

The objectives of the study include the mapping of urban land use change or urban expansion in the past two decades and its impacts on the local environment, including changes in LST, and urban green infrastructure by calculating the NDVI using remote sensing data. This approach to analysis takes into account environemntal sustainability indicators using GIS analysis. The other method of examining of research includes assessing socioeconomic sustainability analysis in the area using multiple linear regression analysis (MLRA).

This approach is based on the statistical investigation of socioeconomic sustainabilty indicators from the study area. These indicators were selected by combing the primary and secondary data, where the primary data consist of pilot surveys and field visits. This approach provided different scenarios of societal and economic conditions in the area. Then a critical analysis was conducted of the Islamabad Master Plan, public opinion regarding socioeconoic conditions, issues highlighted in revised master plans (2020–2040) associated with social factors, the economy, and government documents, that is, Capital Development Authority project reports regarding socioeconomic sustainability, and secondary data were collected.

The secondary phase of the methodology consists of questionnaire design, the calculation of sample size, and completing the questionnaire for statistical analysis. No well-defined method was adopted for measuring socioeconomic sustainability in the past (Niaz, 2022). The major reason for this measurement is the heterogeneous nature of every region or area difference in geography, landscape, environmental values, local characteristics, economic conditions, and social measures. Previous studies, including those by Dong and Hauschild (2017) and Sohail et al. (2023), used economic and social issues for measuring socioeconomic sustainability in an area by statistical analysis such as MLRA. This study involved the local population, stakeholders, and government authorities including the Capital Development Authority, local property dealers, and educators in the survey for data collection to investigate the ground truth and determine the sustainability issues faced by locals. Remote sensing data along with statistical analysis, such as regression analysis, are powerful tools for mapping land use and land cover change and for quantifying socioeconomic conditions through a coefficient of regression and coefficient of correlation among the socioeconomic sustainability parameters (Yu et al., 2020).

The current study used GIS mapping through land use and land cover supervised classification methods on the basis of known samples in the past two decades from the data provided by the United States Geological Survey (USGS), along with the calculation of LST and NDVI as environmental concerns. In addition, MLRA provides a snapshot of current scenarios of land use change and their impacts on economic and societal conditions in the capital city.

The results of the study are the powerful tools for the implementation of the revised master plan of the city and suggest that the government should take measures to shut down illegal housing organizations in the capital and curtail unplanned and rapid migration from the surroundings to the capital, which lead to the conversion of land into urban infrastructure. The study also suggests some other sustainability parameters, including the assessment of air pollution and groundwater potential ratio for ecosystem development to ensure sustainable urban development in the area and make the city a truly sustainable one. Sustainability and the future are closely connected and involve the extensive use of natural resources and urbanization pressure whose effects go beyond the immediate area. These issues pose a potential environmental threat to the entire world particulary, urban areas. Therefore, these kinds of studies are desirable for urban planning and development keeping in mind environmental sustainability.

Literature Review

Sustainability and the future are associated and cause us to focus on conserving resources to meet the demands of future generations, as well as those of the present population. Land use as a component of the environment is an influential sustainability parameter for environmental protection and development in a changing global landscape due to urbanization pressures (Patra et al., 2018). It involves global environmental change and is affected by various factors, including economic changes, housing development, industrialization, urbanization, and population growth. For sustainable development, it is fundamental to estimate spatial changes in a region, such as urban expansion, and its environment and their effect on the local environement. This estimation will lead to increased efficiency of sustainable spatial land use in a changing environment (Dias et al., 2023).

The discussion and examination of land use change are multifactoiral and include forest degradation, agricultural land change, housing, infrastructure development, orchard conversion, crop land conversion, and mixed land use changes. These land use features come in the form of road networks, housing developments, villages, urban areas, commercial areas, zoos, built infrastructure, or mixed infrastructure (Galeano-Barrera et al., 2022). In developing regions, these changes occur rapidly due to increasing demands to meet needs without any concern for the preservation or conservation of resources (James, 2024). This phenomenon will affect spatial sustainability and affect the future in connection with environmental sustainability and development.

The dynamics of land use change are environmental, socioeconomic, and ecological changes and the most critical part of the decision-making process. Therefore, understanding of these environmental, socioeconomic, ecological, and political dynamics can serve as a guide for decision-making with respect to planning and land use development (Bryan et al., 2016).

Multidisciplinary approaches use various methods of studying land use dynamics. Such approaches draw on medicine, engineering, industry, energy, environmental sciences, and urban planning. However, the geographical sciences are more focused on environmental changes and their dynamics including ecosystem services, landscape and geographical changes, and other issues associated with the land in a given environment. These approaches reflect the basic requirements for achieving United Nations Sustainable Development Goals and sustainable urban development.

Research Methodology

Study Area

The selection of the study area is one of the most critical issues in carrying out research on measuring sustainability. This is because every area or region has its own unique needs in terms of its environmental, geographical, social-ecological, and or political contexts (Liu et al., 2023). Islamabad, the capital of Pakistan, was selected as a study area due to its unique geographical location, political importance, planned landscape, and critical need for assessing sustainabily in the region due to urbanization pressures. The area is unique in its infrastructure based on the model presented by Greek architect CA Doxiadis, town and regional planners. Islamabad has a rich urban greenscape thanks to Margalla Hills National Park and mild to moderate climatic conditions. In addition, its inhabitants are highly educated, and it has a well-developed diplomatic infrastructure as the country's capital (Khan, 2006). The most influential factor in the selection of this city in connection with sustainability is its planned infrastructure according to land use zoning based on CA Doxiadis's theory. All of these characteristics make the area the best platform for the estimation of environmental, economic, and social sustainability measures.

In 1960, the city was planned, designed, and built according to Greek town planner and architect Doxiadis or Constantinos Apostolou Dioxide. Its iron gridlike structure gives it a beautiful landscape and highlights it as a well-planned area (Authority, 2020). The map in Fig. 1 shows the location of the study area.

The city is located in the northern Punjab at the foot of the Margalla Hills (forested area at the northern edge of the city), at the edge of the Pothohar Plateau neighboring the Rawalpindi district of Punjab, Murree in the east, and the Haripur district of Khyber Pakhtunkhwa (KPK). The city's elevation is 507 m, with a total aera of 906.5 km^2.

The city of Islamabad has been undergoing rapid urbanization since the start of the twenty-first century due to the expansion of urban infrastructure, settlements in the city limits, road networking, and industrial growth (Liu et al., 2021; Wahab et al., 2023). On the other hand, the surrounding areas have also merged with the city limits, which indirectly enlarged the magnitude of urban land use and expanded urbanization. As a result, over the past two decades, the city has been facinng urbanization pressure, and the area is struggling to contain the population boom, according to the National Bureau of Statistics Pakistan (Pakistan, 2017). This increasing population has caused unplanned city planning and

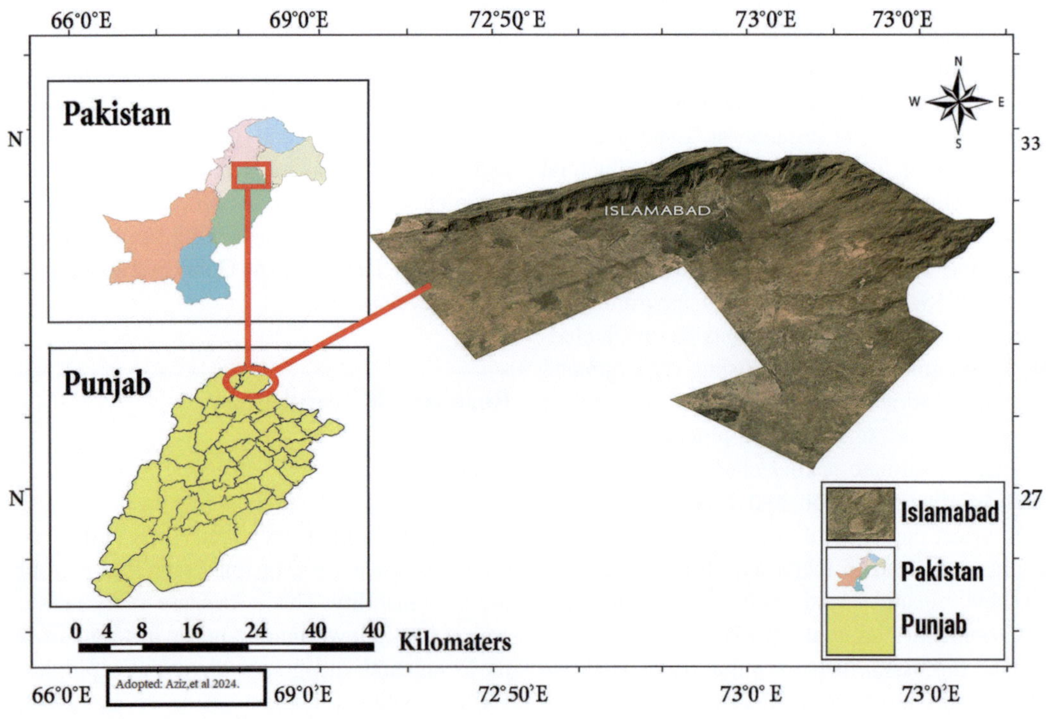

Fig. 1 Map of study area, Islamabad, Pakistan

development and challenged urban sustainability in the area, which needs to be addressed to ensure a better future for the region.

The increasing urban infrastructure due to urban land use expansion also increased the local atmospheric temperature and the conversion of forest cover to a built environment in terms of biodiversity loss of forest. Addintionally, the increasing population as a result of migration from surrounding areas also affects the local economic and social conditions by increasing the demand for jobs, natural resources, and social services, including social security, housing, commercial space for economic activity, and residential areas. The current scenarios of urban expansion may lead to the growing sprawl of city areas. This will lead to multiple issues for regional development and sustainability and will need to be addressed (Sohail et al., 2019; Tariq & Waheed, 2023).

Materials and Methods

Because of sustainability's diverse nature and varying definitions, it is very difficult to measure. No well-defined method has been introduced to measure the sustainability of an area using any particular objective or parameters (Boggia & Cortina, 2010). This is because every area, city, or region at a different geographical scale has its own needs or demands for sustainability due to diverse area characteristics and varying regional geographical landscapes. However, sustainability is more concerned with the environment, ecosystem services, and economic and social means in a society (Mura et al., 2021).

This study used mixed methods approaches, including qualitative and quantitative analysis using GIS and multiple regression analysis for environmental, economic, and social measurements of sustainability. These approaches included GIS mapping as a qualitative approach using supervised classification methods of unknown pixels and constructing different classes for every major land use. On the other hand, the quantitative approach was based on statistical methods via MLRA. The coefficient of correlation (R) and coefficient of regression were used to express the results of the socioeconomic sustainabiloty variables used.

Data Analysis

Regression was applied on the secondary data collected from completed questionnaires using a 5-point Likert scale on dependent and independent variables. This analysis is concerned with economic and social value measurements and their relation to sustainability through different parameters. This part of the methods section consists of a pilot survey of the study area, ground truthing, review of past studies, review studies, selection of variables for socioeconomic sustainability analysis, questionnaire and sample design, corn bench alpha analysis, and data collection. Corn bench analysis provides the internal consistency of used variables to avoid bias in the results. Further, this section also includes a data analysis by MLRA in Statistical Package for the Social Sciences (SPSS). This section provides the coefficient of correlation (R) and coefficient of regression (R^2). Additionaly, the detailed analysis is described in the next phases of the methodology.

Environmental Sustainability Measures

Environmental Sustainability, Urban Expansion, and Associated Data

The environmental parameters of sustainability including land use and land cover change, in terms of urban expansion, total variation in vegetation index, and land surface change, were quantified using GIS analysis. Remote sensing data were collected from an official source, the USGS, as detailed in Table 1.

Remote sensing data were collected from the USGS from the past two decades. These data were further analyzed by pixel-based classification in ArcMap for the assessment of land use and land cover change or urban expansion in the study area by constructing various classes—

Table 1 GIS data sources

Year	Date obtained	Product	Numbers of bands	Pixel size (m)	Path and row	Cloud cover (%)
2000	September 13, 2023	Landsat 5	7	30	150/037	8
2010	September 4, 2023	Landsat 7	8	30	150/037	6
2015	October 15, 2023	Landsat 8	9	30	150/037	7
2020	October 18, 2023	Landsat 8	9	30	150/037	8

built-up infrastructure, water bodies, barren land, and vegetation. Furthermore, the spatial analysis tool was used for other environmental sustainability variables like LST and NDVI. ArcMap 10.8 was used for the calculation of the NDVI and LST for the past two decades.

Socioeconomic Sustainability Data Collection

Socioeconomic sustainability data availability, collection, examination, and presentation are difficult. For the collection of these data, a pilot survey was conducted to check the ground truthing and investigate the issues regarding socioeconomic conditions challenging sustainability in the study area. This socioeconomic survey was done in the urban area of Islamabad as a basis for discussing with locals economic and social issues by random sampling methods. These questions were of a general nature, associated with economic sustainability, and related to the availability of jobs in the area, land values and housing rent, employment rate per household, and household size of the respondents. Similarly for social sustainability measures, the questions related to social security in a residential community (SSRC), soil waste management (SWM), and provision of neighborhood services (PNS). For greater clarity on the concept, a closed-ended questionnaire was filled out by experts. These experts included environmentalists, urban property dealers, housing society managers, social development officers, municipal authorities, and officials from Capital Development Authorities. This field survey through random sampling provides comprehensive data regarding socioeconomic sustainability issues and their parameters to design a questionnaire for data collection and analysis. The major resources for data collection variables was the revised master plan (2020–2040), urban property dealer's records, urban gazette, published reports by the research journals, and government records. A 5-point Likert scale ranging from 1, for "strongly agree," to 5, for "strongly disagree" was used to collect the responses from the respondents. In the Likert scale, urban sustainabilty was the dependent variable, while socioeconomic sustainability variables were independent variables.

Analyses

The analysis of the study consists of two parts: GIS analysis by ArcMap 10.8 using remote sensing data and Statistical Package for the Social Sciences (SPSS) to examine socioeconomic sustainability via regression analysis. Land use classification was done by taking the points of known topographical features (known samples) for land use classes, including built-up, water bodies, vegetation, and barren land. On the other hand, band 5 and spectral radiance were used to measure the LST. The vegetation index values, ranging from −1 to +1, were calculated by the GIS analysis of bands 3, 4, and 5 and Land-sat 8 and 9 of the past two decades. Equation (1) was used for these calculations:

$$\mathrm{NDVI} = \left(\mathrm{NIR} - R\right) / \left(\mathrm{NIR} + R\right) \tag{1}$$

The second part of the methodology is mainly concerned with the social analysis of economic and social parameters of sustainability. Multiple regression analysis techniques were applied to examine the data collected by the questionnaire, and these data were treated as secondary data.

Selection of Variables and Regression Analysis

There are no well-defined methods or data available to measure socioeconomic sustainability, particularly in developing states. The selected socioeconomic sustainability questionnaire for understanding public opinion included questions on rent of commercial spaces (RCS), property rent (PR), land values in residential areas (LVRA), economic opportunity (EO), social security in a residential community (SSRC), soil waste management (SWM), and provision of neighborhood services (PNS).

$$\text{Urban sustainability}_{(\text{Predicted})} = \left(\text{Economic parameters} + \text{Social parameters}\right) \quad (2)$$

In Eq. (2), the socioeconomic sustainability variables are independent variables and urban sustainability is a dependent variable for regression analysis in SPSS. The next equation, Eq. (3), provides the regression formula for this estimation as a regression equation used for statistical analysis in SPSS:

$$\Sigma \text{US} = \Sigma \beta_O + \beta_{\text{RCS}} X_{\text{RCS}} + \beta_{\text{PR}} X_{\text{PR}} + \beta_{\text{LVRA}} X_{\text{LVRA}} + \beta_{\text{EO}} X_{\text{EO}} + \beta_{\text{SWM}} X_{\text{SWM}} + \beta_{\text{SSPS}} X_{\text{SSPS}} + u, \quad (3)$$

where US is urban sustainability that is a predicted dependent variable in regression analysis; it is the sum of socioeconomic sustainability parameters including RCS, PR, LVRA, EO, SSRC, SWM, and PNS; β_O is the intercept; and u is the model error. *Multiple linear regressions were used to quantify the impacts of these variables on urban sustainability levels in the area. In statistical analysis, the results of the variables used for measuring urban sustainability were examined. The coefficient of regression is represented by R, and the coefficient of determination is represented by* R^2.

Previous research found that there no specific methods existed to measure socioeconomic sustainability using fixed variables or parameters (Denu et al., 2023; Vargas-Santander et al., 2023; Woodcraft, 2015). All regions have their own particular economic means and social values in terms of their morphological characteristics and magnitude. In the case of developing states including Pakistan, the lack of planning also challenges the socioeconomic properties of the region even at the local level (Gilani et al., 2020).

However, this study involved the local community, expert opinions, past research, and government documents to select the variables and methods to examine sustainability in the capital of Pakistan. To avoid bias, the study obtained its primary data only from educated locals, and a total of 400 questionnaires were completed after establishing the sample. The sample size was calculated through an online source using a sampling calculator (https://www.calculator.net/sample-size-calculator.html). For validation of data regarding land values, the property records of local property dealers were also checked. Finally, the results obtained from the analysis were compared with other official reports, past research associated with development plans, and sustainability for validation and accuracy.

Results

Environmental Sustainability and GIS Results

Environmental sustainability includes such environmental parameters as land use change in the past two decades and total change in vegetation and LST.

Figure 2 provides the GIS mapping analysis of land use and land cover in the past two decades. Table 2 provides the information total change that occurred in the built-up area, water bodies, barren land, and vegetation. Significant change occurred in urban built-up areas and barren land.

In the year 2000, the settled area was only 23.57 km^2, which has increased to 312.35 km^2

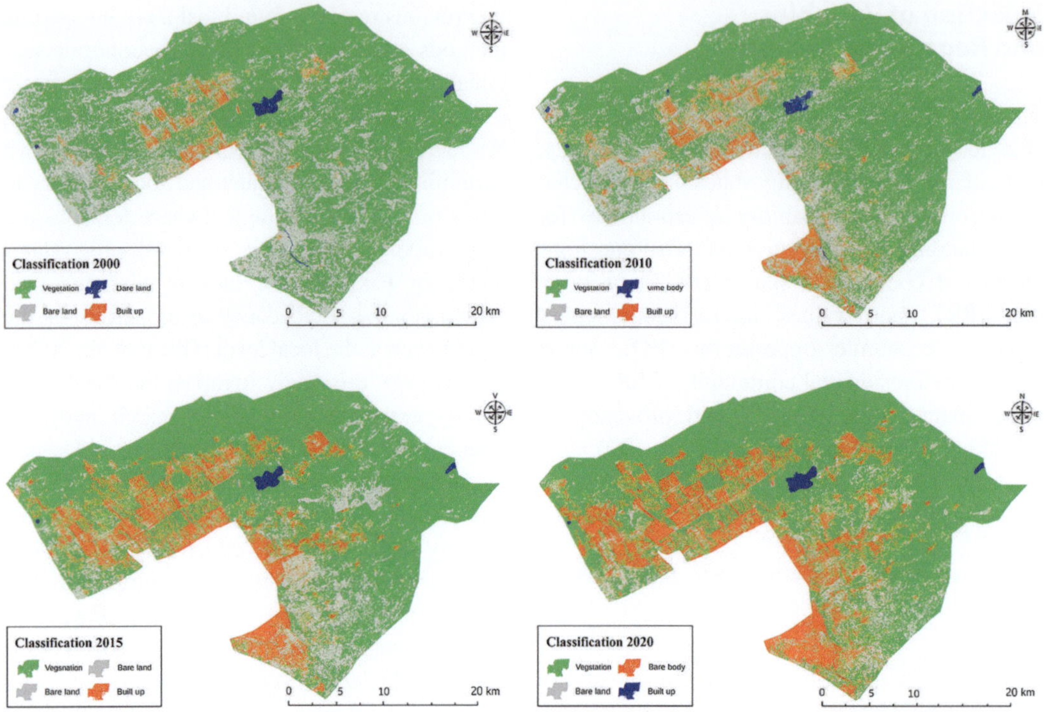

Fig. 2 Land use classification map of Islamabad, Pakistan

Table 2 Statistical results of land use dynamics in Islamabad, Pakistan

Dynamics of land use and land cover change	Change in land use values in 2000, 2010, 2015, and 2020 in Islamabad, Pakistan			
	2000	2010	2015	2020
	km^2 and (%)	km^2 and (%)	km^2 and (%)	km^2 and (%)
Water	7.7 (0.887%)	8.28 (0.89%)	9.43 (0.98%)	8.87 (1.20%)
Vegetation	517.55 (56.54%)	438.66 (55.69%)	420.09 (58.42%)	380.00 (44.28%)
Barren land	344.09 (36.93%)	323.46 (36.61%)	233.61 (27.25%)	212.68 (24.94%)
Built up	23.57 (3.571%)	66.62 (9.38%)	144.65 (16.96%)	312.35 (36.76%)

during the last 20 years, which is evidence of urban expansion. Similarly, water bodies have increased from 7.7 to 8.87 km^2 in the last two decades.

This increase is mainly caused by the construction of dams by the government. Meanwhile, in the case of barren land, housing societies cover the area with colonies and housing networks for rent and other commercial purposes, which decreased this area from 344.09 to 212.68 km^2. The vegetation area was also affected by urbanization and built-up infrastructure and decreased from 517.55 to 380 km^2.

For validation of the GIS analysis of land use classification, an accuracy assessment was done to ensure the reliability of the spatial analysis. In the GIS environment, the spatial user and producer accuracy was obtained by running an overall accuracy matrix. For user accuracy, the total classification was divided by the total reference points taken from the Google Earth Pro version. Meanwhile, the producer accuracy results were obtained by dividing the classified land use and land cover cells with total ground truth pixels. The validation of GIS analysis is a very important part of landscape planning and the spatial sustainability of an area. Finally, the overall accuracy was calculated by dividing the classified cells by a total number of pixels, and the obtained results are presented in Table 3.

Table 3 Accuracy assessment of LULC change from (2000 to 2020)

Class name	User accuracy (%)	Producer accuracy (%)	Overall accuracy
Year 2000			
Water cover	92.34	91.67	0.87
Barren land	95.84	89.56	
Vegetation	88.45	81.56	
Built up area	92.34	83.67	
Year 2020			
Water cover	83.12	72.56	0.82
Barren land	79.43	76.45	
Vegetation	86.45	81.34	
Built-up area	81.70	72.50	

The second parameter for measuring environmental sustainability is the LST, which also significantly increased due to human activities, urbanization, and removal of urban green infrastructure in the city. The map in Fig. 3 provides the LST variation from 2000 to 2020.

Similarly, the third parameter of environmental sustainability is the vegetation index in the city, which was measured by the NDVI. From 2000 to 2010 a significant decrease in vegetation was noted due to forest cover lost. From 2011 to 2020 there was an increase in the vegetation index due to the efforts of the government to clean the environment by increasing green spaces in the city through plantations, urban parks, and a ban on deforestation in the Margalla Hills. Figure 4 illustrates this variation in the NDVI map of Islamabad, Pakistan.

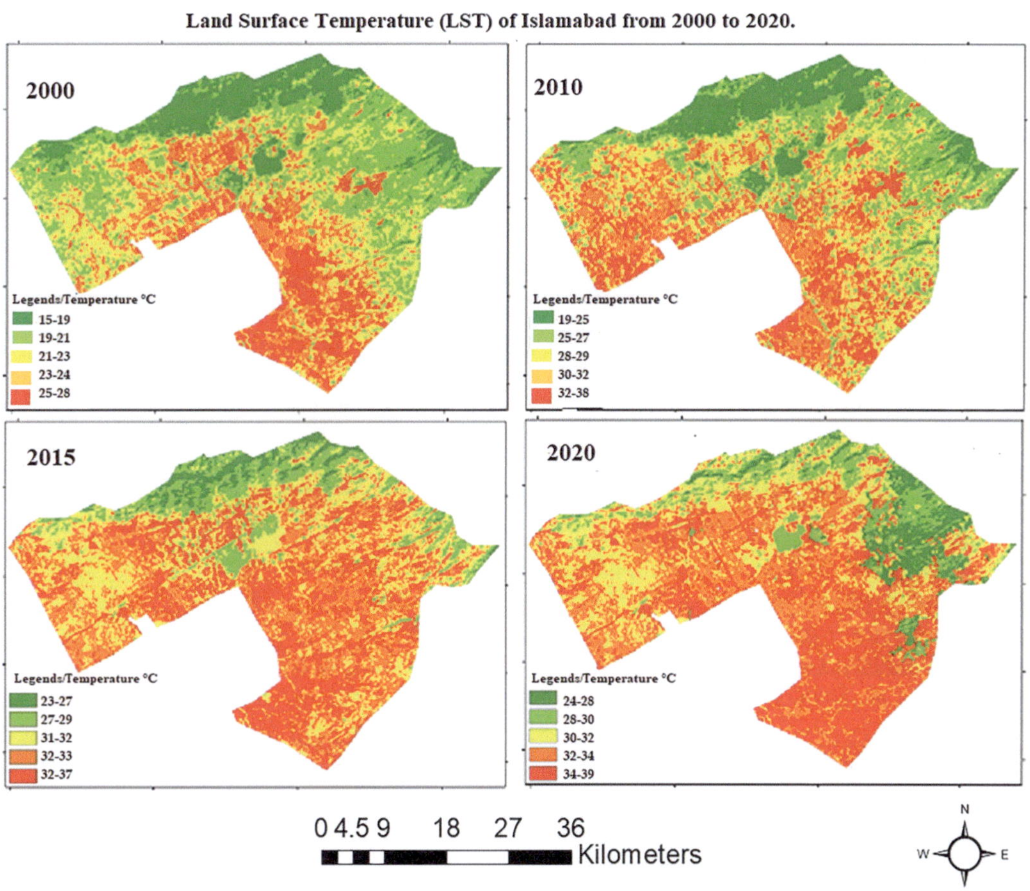

Fig. 3 Land surface temperature of Islamabad

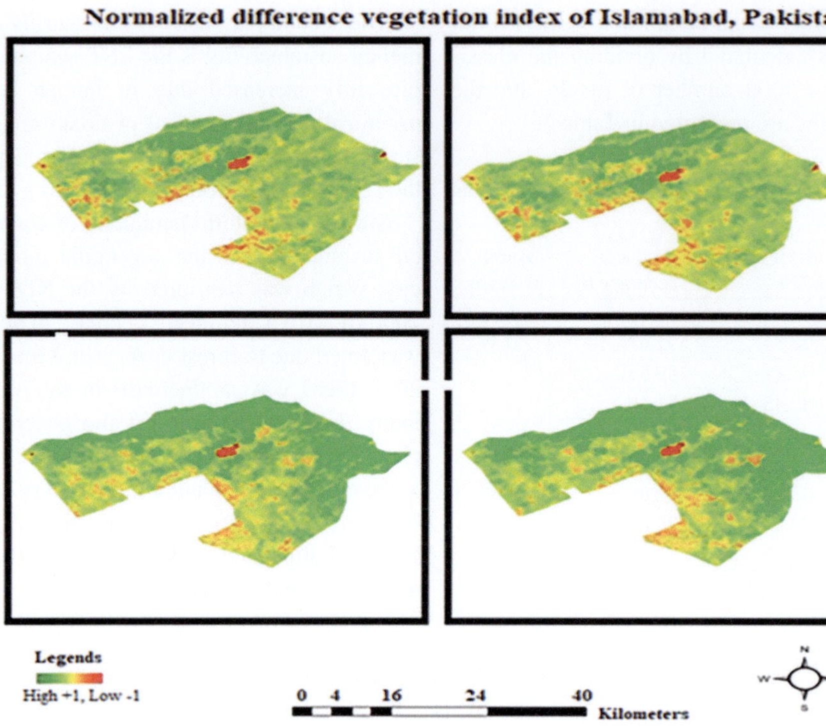

Fig. 4 NDVI map of Islamabad, Pakistan

Socioeconomic Sustainability

To quantify socioeconomic sustainability, the primary data provide comprehensive information and a road map regarding economic and social issues, including in the area discussed earlier; then the secondary data are used for regression analysis. The coefficients of regression show a strong correlation between the selected socioeconomic parameters, for example, land values, housing prices, property rent, social security, sense of belonging, provision of social services, and availability of jobs in the area. *To check the internal consistency and avoid inaccuracy and biases, a reliability test, Cronbach's alpha test in* Table 4, *was run and the results obtained were as follows.*

Table 4 Reliability test for socioeconomic sustainability indicators (overall)

Cronbach's alpha	No. of items
0.774	07

Table 5 Reliability test for economic sustainability indicators (individual)

Indicator	Cronbach's alpha values
Commercial space rent	0.784
Property rent	0.872
Land values in area	0.789
Economic opportunities	0.632
Social security	0.872
Solid waste management	0.783
Size of houses	0.892

In addition to this, the reliability test was run individually for each selected indicator, and the results obtained were as shown in Table 5.

Table 6 shows the correlation and regression results through the values of R and R^2 taken as the coefficient of regression and coefficient of correlation. The values of R indicate that the selected variables define the level of sustainability by 79% in the correct order. This means that the overall test reliability is good and the study analysis demonstrates that the socioeconomic sustainability depends on these variables by 80% in the area. In addition to this, our results reveal that the high

Table 6 Regression statistics of socioeconomic sustainability in Islamabad

R	R^2	Adjusted R^2
0.791	0.731	0.801

economic values of land and property also affect the rent of space for commercial and residential activities, which poses a challenge to economic sustainability in the area. The rising growth rate of the population also requires more resources, ultimately increasing demand for social and economic services and damaging the existing infrastructure of local sustainability efforts. Therefore, a complete follow-up of the revised master plan (2020–2040) urgently requires an area for society and the sustainable development of that area.

Discussion

Sustainability and sustainable development are associated, but there are no well-defined methods to measure them. Due to geological, geographical, social, political, and economic heterogeneity, it is difficult to measure or estimate sustainability and sustainable development. This study made an attempt to quantify the level of urban sustainability in Pakistan's capital through environmental, economic, and social parameters. These parameters includes the economic opportunities, social peace, political and physical environmental stability, and their services. The results of the study depicts an alamring situation in the area with respect to stability in environmental, economic, and societal terms. This stability should be maintained over time without exceeding the carrying capacity of the environment to ensure the urban sustainability of the area (Neuman & Churchill, 2015). Meanwhile, in developing countries, including Pakistan, the carrying capacity is at risk due to overexploitation of natural resources and ecosystems, which threatens sustainability, particularly in major urban areas. The same situation has been observed in the study area of Islamabad, Pakistan, which needs to be addressed to protect the city's spatial morphology and land use for sustainable development.

As a result of the study's findings, it is recommended that the government's Capital Development Authority ban deforestation in Margalla Hills and the illegal housing organizations that operate in the city, along with their housing schemes, be shut down to protect the socioeconomic integrity of the area. The quantitative sustainable development parameters of our study provide comprehensive data regarding the socioeconomic sustainability of the capital of Pakistan, which will be helpful for area development and socioeconomic development of the area.

Conclusions

Sustainability and the future are correlated. Urbanization pressure represents one of the gravest threats for the future and sustainability. The threat lies in the conversion of land and forest into built-up infrastructure, which spurs increases in temperature in the region, which in turn causes significant changes in local weather conditions. The conversion of urban green forests into road networks and housing developments alters the local ecosystem and challenges the socioeconomic sustainability of the area. It also causes an increase in security threats, resource depletion, and deficiencies in neighborhood services in the area.

During the past two decades, the total change in land use dynamics, including water, from 7.7 to 9.43 m² shows an increase of water resource use in the city made possible by dams. The total change in built-up area is 23.57–312.35 m² mainly due to an increase in urban expansion and residential housing, even for barren land. In the case of vegetation cover, the total change is a 480-m² decrease due to road networking and other construction projects. The increase in population also spurs the daily demand for economic, social, and environmental services, but the area is unable to provide these services. The GIS analysis that was conducted, which included LST and land cover maps, depicts a very high rate of urbanization in this area. This high rate of urban expansion degrades the environment and the local ecosystem and creates pollution. The high population requires more job opportunities, which will pose a challenge for economic development in the area. The faster rate of migration changes the demographic characteristics of the city, which affects social cohesion, diversity, sense of belonging, and social integrity.

Smart technology and innovations such as liquid trees and trams that use artificial intelligence in populated regions, with the help of big data, are the most effective tools for achieving the goal of sustainable urban development. Different smart technologies like liquid trees should be installed in congested areas of the city to decrease carbon emissions and reduce pollution. Similarly, trams or smart Bus Rapid Transit (BRT) transportation system should be provided to the public to increase accessibility and economic sustainability.

The study concluded that the Capital Development Authority should refine and redefine policies for new construction and allocate land to new residences outside the major urban area. The expansion of existing political boundaries is a fundamental need of the city to cope with the needs of the city's population. The private sector should coordinate with the government for the provision of socioeconomic services for area sustainability. Finally, it is essential to adopt and follow the revised master plan of the city to transform it into a truly sustainable city.

References

Arfanuzzaman, M., & Dahiya, B. (2019). Sustainable urbanization in Southeast Asia and beyond: Challenges of population growth, land use change, and environmental health. *Growth and Change, 50*, 12297. https://doi.org/10.1111/grow.12297

Authority. (2020). *Revised master plan of Islamabad*. The Government of Pakistan. Retrieved from https://www.cda.gov.pk/about_islamabad/mpi/reports

Benjamin, S. (2015). *Cities within and beyond the plan*. Routledge.

Boggia, A., & Cortina, C. (2010). Measuring sustainable development using a multi-criteria model: A case study. *Journal of Environmental Management, 91*(11), 2301–2306. https://doi.org/10.1016/j.jenvman.2010.06.009

Bryan, B. A., Nolan, M., McKellar, L., Connor, J. D., Newth, D., Harwood, T., King, D., Navarro, J., Cai, Y., Gao, L., Grundy, M., Graham, P., Ernst, A., Dunstall, S., Stock, F., Brinsmead, T., Harman, I., Grigg, N. J., Battaglia, M., & Hatfield-Dodds, S. (2016). Land-use and sustainability under intersecting global change and domestic policy scenarios: Trajectories for Australia to 2050. *Global Environmental Change, 38*, 130–152. https://doi.org/10.1016/j.gloenvcha.2016.03.002

Denu, M. K., Bentley, Y., & Duan, Y. (2023). Social sustainability performance: Developing and validating measures in the context of emerging African economies. *Journal of Cleaner Production, 412*, 137391. https://doi.org/10.1016/j.jclepro.2023.137391

Dias, F. T., Mazon, G., Cembranel, P., Birch, R., & de Andrade Guerra, J. B. (2023). Land use and global environmental change: An analytical proposal based on a systematic review. *Land, 12*(1), 115.

Dong, Y., & Hauschild, M. Z. (2017). Indicators for environmental sustainability. *Procedia CIRP, 61*, 697–702. https://doi.org/10.1016/j.procir.2016.11.173

Galeano-Barrera, C. J., Arango Ospina, M. E., Mendoza García, E. M., Rico-Bautista, D., & Romero-Riaño, E. (2022). Exploring the evolution of the topics and research fields of territorial development from a comprehensive bibliometric analysis. *Sustainability, 14*(11), 6515.

Gilani, H., Ahmad, S., Qazi, W. A., Abubakar, S. M., & Khalid, M. (2020). Monitoring of urban landscape ecology dynamics of Islamabad Capital Territory (ICT), Pakistan, over four decades (1976–2016). *Land, 9*(4), 123.

James, N. (2024). Urbanization and its impact on environmental sustainability. *Journal of Applied Geographical Studies, 3*, 54–66. https://doi.org/10.47941/jags.1624

Kalfas, D., Kalogiannidis, S., Chatzitheodoridis, F., & Toska, E. (2023). Urbanization and land use planning for achieving the sustainable development goals (SDGs): A case study of Greece. *Urban Science, 7*(2), 43.

Khan, A. (2006). Dynapolis and the cultural aftershocks: The development in making of Islamabad and the reality today. *Journal of Research in Architecture and Planning, 5*(1), 1–20. Retrieved from https://difusion.ulb.ac.be/vufind/Record/ULB-DIPOT:oai:dipot.ulb.ac.be:2013/154660/Details

Li, Y., Zhang, X., Zhao, X., Ma, S., Cao, H., & Cao, J. (2016). Assessing spatial vulnerability from rapid urbanization to inform coastal urban regional planning. *Ocean & Coastal Management, 123*, 53–65. https://doi.org/10.1016/j.ocecoaman.2016.01.010

Liu, F., Dai, E., & Yin, J. (2023). A review of social–ecological system research and geographical applications. *Sustainability, 15*(8), 6930.

Liu, Y., Rahimoon, S., & Jiang, Y. (2021). Urban growth sustainability of Islamabad, Pakistan, over the last 3 decades: A perspective based on object-based backdating change detection. *GeoJournal, 86*, 2035–2055. https://doi.org/10.1007/s10708-020-10172-w

Mahadevia, D., Mukhopadhyay, C., Lathia, S., & Gounder, K. (2023). The role of urban transport in delivering sustainable development goal 11: Learning from two Indian cities. *Heliyon, 9*(9), e19453. https://doi.org/10.1016/j.heliyon.2023.e19453

Mallick, S. K., Das, P., Maity, B., Rudra, S., Pramanik, M., Pradhan, B., & Sahana, M. (2021). Understanding future urban growth, urban resilience and sustainable development of small cities using prediction-adaptation-resilience (PAR) approach. *Sustainable Cities and Society, 74*, 103196. https://doi.org/10.1016/j.scs.2021.103196

Mura, M., Longo, M., Toschi, L., Zanni, S., Visani, F., & Bianconcini, S. (2021). The role of geographical scales in sustainability transitions: An empirical investigation of the European industrial context. *Ecological Economics, 183*, 106968. https://doi.org/10.1016/j.ecolecon.2021.106968

Neuman, M., & Churchill, S. (2015). Measuring sustainability. *Town Planning Review, 86*, 457–482. https://doi.org/10.3828/tpr.2015.28

Niaz, M. U. (2022). Socio-economic development and sustainable development goals: A roadmap from vulnerability to sustainability through financial inclusion. *Economic Research-Ekonomska Istraživanja, 35*(1), 3243–3275. https://doi.org/10.1080/1331677X.2021.1989319

Pakistan. (2017). *Population census 2017*. Pakistan Bureau of Statistics. Retrieved from https://www.pbs.gov.pk/content/population-census

Pandey, A., & Asif, M. (2022). Assessment of energy and environmental sustainability in South Asia in the perspective of the sustainable development goals. *Renewable and Sustainable Energy Reviews, 165*, 112492. https://doi.org/10.1016/j.rser.2022.112492

Patra, S., Sahoo, S., Mishra, P., & Mahapatra, S. C. (2018). Impacts of urbanization on land use/cover changes and its probable implications on local climate and groundwater level. *Journal of Urban Management, 7*(2), 70–84. https://doi.org/10.1016/j.jum.2018.04.006

Roy, A. (2012). Urban informality: The production of space and practice of planning. In *The Oxford handbook of urban planning*. Oxford University Press. https://doi.org/10.1093/oxfordhb/9780195374995.013.0033

Sohail, D., Mahfooz, Y., Azam, A. P. D. K., Yat, Y., & fahad, s. (2019). Impacts of urbanization and land cover dynamics on underground water in Islamabad, Pakistan. *Desalination and Water Treatment, 159*, 402–411. https://doi.org/10.5004/dwt.2019.24156

Sohail, M. T., Manzoor, Z., Ehsan, M., Al-Ansari, N., Khan, M. B., Shafi, A., Ullah, J., Hussain, A., Raza, D., Usman, U., Akbar, S., & Elbeltagi, A. (2023). Impacts of urbanization, LULC, LST, and NDVI changes on the static water table with possible solutions and water policy discussions: A case from Islamabad, Pakistan. *Frontiers in Environmental Science, 11*, 1018500. https://doi.org/10.3389/fenvs.2023.1018500

Tariq, A., & Waheed, A. (2023). Urbanization and urban patterns: The Islamabad experience. *Journal of Xi'an Shiyou University, 19*(12), 326–350. Retrieved from https://www.xisdxjxsu.asia/V19I12-19.pdf

Vargas-Santander, K. G., Álvarez-Diez, S., Baixauli-Soler, S., & Belda-Ruiz, M. (2023). Developing a country's sustainability indicator: An analysis of the effect on trade openness. *Environmental and Sustainability Indicators, 19*, 100280. https://doi.org/10.1016/j.indic.2023.100280

Wahab, Z., Wahab, A., Farah, H., Waheed, A., & Sahbeni, G. (2023). Examination of spatio-temporal urbanization patterns in Islamabad Metropolis, Pakistan, over past four decades: A remote sensing–based approach. *Arabian Journal of Geosciences, 16*, 62. https://doi.org/10.1007/s12517-022-11121-2

Woodcraft, S. (2015). Understanding and measuring social sustainability. *Journal of Urban Regeneration and Renewal, 8*, 133–144.

Yu, H., Song, G., Li, T., & Liu, Y. (2020). Spatial pattern characteristics and influencing factors of green use efficiency of urban construction land in Jilin Province. *Complexity, 2020*, 5637530. https://doi.org/10.1155/2020/5637530

Zhou, G., Zhang, J., Li, C., & Liu, Y. (2022). Spatial pattern of functional urban land conversion and expansion under rapid urbanization: A case study of Changchun, China. *Land, 11*(1), 119.

The Role of Small Public Green Spaces in Regulating Urban Microclimate: Insights from Bucharest

Gabriel Ovidiu Vânău,
Athanasios-Alexandru Gavrilidis,
Diana Andreea Onose, Ana-Maria Popa,
and Raluca Andreea Slave

Abstract

Urbanization and climate change have exacerbated the urban heat island (UHI) effect, prompting an urgent need for sustainable, space-efficient cooling solutions. This study investigates the role of small public urban green spaces (SPUGS) in regulating urban microclimates, particularly within Bucharest's densely populated neighborhoods. Although extensive research has focused on large green spaces, SPUGS may offer a viable alternative, especially in cities where large green areas are scarce. This research hypothesizes that SPUGS significantly impact temperature regulation, providing a cumulative cooling effect within urban fabrics. Using a UAV equipped with temperature data loggers, vertical and horizontal air temperature measurements were recorded across a designated neighborhood in Bucharest. Temperature data was collected at multiple altitudinal thresholds, revealing significant temperature reductions around green spaces, particularly at ground level and just above tree canopies. Spatial interpolation and correlation analyses showed a strong negative relationship between vegetation coverage and temperature variance, confirming that higher vegetation density corresponds to lower temperature differentials. These findings highlight the utility of SPUGS in urban planning for their cooling potential, ease of implementation, and ability to enhance microclimatic resilience.

Keywords

UHI · Microclimate regulation · Climate changes · UAV-based temperature analysis

G. O. Vânău · A.-A. Gavrilidis (✉) ·
D. A. Onose · A.-M. Popa
Department of Regional Geography and Environment, University of Bucharest, Bucharest, Romania

Centre for Environmental Research and Impact Studies, University of Bucharest, Bucharest, Romania
e-mail: gabriel.vanau@geo.unibuc.ro; athanasiosalexandru.gavrilidis@g.unibuc.ro; dianaandreea.onose@g.unibuc.ro; anamaria.popa@geo.unibuc.ro

R. A. Slave
Centre for Environmental Research and Impact Studies, University of Bucharest, Bucharest, Romania

Doctoral School – "Simion Mehedinți", University of Bucharest, Bucharest, Romania
e-mail: andreea.slave@drd.unibuc.ro

M. M. Anwar, A. Aziz (eds.), *Sustainable Landscape for Urban Resilience*, Sustainable Development Goals Series, https://doi.org/10.1007/978-3-031-99836-2_3

Introduction

Enhancing resilience in urban systems requires the development of frameworks that enable cities to withstand, recover from, and adapt to environmental stresses. A resilience-focused approach is critical, prioritizing improvements in essential infrastructure—such as water, energy, and transportation networks—to ensure functionality during extreme events and to support both immediate recovery and sustained adaptation (Ribeiro & Gonçalves, 2019). The rapid pace of urbanization, combined with emissions from transportation and industrial activities, has intensified the urban heat island (UHI) effects, wherein urban areas experience significantly warmer temperatures than nearby rural zones. This warming effect contributes to the formation of "*urban pollution islands*," where temperature-dependent chemical reactions raise the values of ground-level ozone and other pollutants (Ulpiani, 2021). Additionally, the expansion of peri-urban areas has led to low-density developments that strain urban infrastructure, complicating the consistent delivery of essential services such as water supply, transportation, and waste management across both established urban centers and newly developed areas (Shaw et al., 2020). Urban critical infrastructure networks are increasingly vulnerable to both natural and human-induced hazards. Disruptions in these networks can result in substantial economic losses and jeopardize public safety (Liu & Song, 2020). To mitigate these challenges, integrating green infrastructure (GI) and nature-based solutions (NBS) offers sustainable and multifunctional approaches. Green spaces and vegetative cover not only reduce UHI effects and improve air quality, but also support biodiversity. As such, GI and NBS have become central strategies for advancing urban sustainability, addressing urban challenges like heat waves, flooding, and pollution in an effective, multifunctional manner (Fang et al., 2023).

Urban green spaces are instrumental in addressing environmental challenges in cities, particularly as urbanization and climate change intensify pressures on urban areas. These green spaces enhance urban resilience by providing essential regulatory ecosystem services, which help mitigate adverse environmental impacts. For example, they play a role in filtering air pollution by capturing airborne particulates, contribute to local climate regulation by cooling urban temperatures, and manage stormwater, reducing flood risks (Fang et al., 2023; Kabisch et al., 2017). Furthermore, urban green spaces mitigate the UHI effect, cooling surrounding areas through shading and evapotranspiration, thereby lowering energy demands for air conditioning and improving urban livability (Evans et al., 2022).

In addition to these regulatory functions, urban green spaces offer a range of ecosystem services that improve quality of life. Provisioning services, such as urban agriculture and community gardens, can support local food security, while cultural services provide residents with areas for recreation, social interaction, and mental health benefits (Evans et al., 2022; Pandey & Ghosh, 2023). These benefits are especially valuable for vulnerable groups, such as the elderly and children, who gain from accessible green spaces that encourage physical activity, reduce stress, and foster community cohesion (Gavrilidis, Zakerhaghighi, et al., 2023; Zhang & Qian, 2024). In offering these diverse benefits, green spaces support sustainable development goals, promote biodiversity, and strengthen urban resilience against climate-related stressors, making their integration into urban planning increasingly essential for sustainability and public well-being (Fang et al., 2023).

Shading from trees can reduce surface and air temperatures by up to 3 °C, particularly benefiting urban residents by alleviating heat stress during high-temperature periods (Ettinger et al., 2024; Marando et al., 2019). Moreover, the cooling effects of green spaces extend beyond their immediate location, creating a "*cooling halo*" that benefits nearby built environments.

The structure and distribution of green spaces are critical to maximizing their temperature-regulating capacity. Larger, continuous green areas and densely vegetated spaces are particularly effective in lowering land surface temperatures, while fragmented or sparsely vegetated spaces offer less cooling potential. Nonetheless, strategically distributed smaller green spaces can form a "*patchwork*" cooling network, improving microclimates across densely populated areas that lack large parks (Egerer et al., 2024; Fotoohi et al., 2024).

Through these varied mechanisms, urban green spaces provide crucial thermal regulation services, emphasizing their value as a climate adaptation strategy that enhances urban resilience against rising temperatures and extreme heat events. Consequently, urban planners are increasingly incorporating extensive green infrastructure, particularly in densely built areas that are most vulnerable to heat retention (Bai et al., 2024).

Small public urban green spaces (hereafter referred to as SPUGS) play a crucial role in temperature regulation, particularly in urban environments facing the challenges of climate change and increased population density. Studies have shown that small green patches dispersed throughout urban areas contribute to a cumulative cooling impact, which benefits local microclimates, particularly in densely built neighborhoods where green space is otherwise scarce (Gavrilidis et al., 2022).

Though most researchers focused their attention on urban forests, urban parks, or other larger green features, SPUGS may prove to be the assets required to address climate change adaptation in large cities as land availability for large-sized green areas is scarce. Moreover, the distribution of large urban green areas within a city tends to favor affluent and developed neighborhoods. Previous studies have highlighted disparities in access to green spaces, emphasizing how unequal distribution impacts urban sustainability and social justice, particularly for lower-income residents (Anwar et al., 2023).

Analyses at finer spatial scales, such as neighborhood levels, reveal disparities in park provision that regional analyses might overlook, emphasizing the need for small-scale, localized planning to address community-specific accessibility challenges (Tan & Samsudin, 2017). Hence, various methods have been tested for assessing SPUGS such as Geographic Object-Based Image Analysis (GEOBIA) and high-resolution remote sensing—increasingly used to map and monitor small green areas accurately or vegetation indices such as the normalized difference vegetation index (NDVI) and modified soil adjusted vegetation index (MSAVI)—commonly employed to quantify vegetation health, density, and distribution, providing insights into the quality and impact of small-sized green areas across diverse urban climates (Popa et al., 2022). Studies addressing the role of green areas in mitigating air pollutants concentration in cities revealed that urban green spaces consistently improve air quality, but the effects vary significantly based on green space type, size, and biome. However, tree cover was linked to reductions in pollutants like O_3, PM_{10}, and $PM_{2.5}$, while at street level, dense vegetation within street canyons sometimes exacerbated pollution by limiting airflow and increasing pollutant concentration (Venter et al., 2024).

Beyond temperature regulation, small urban green spaces provide essential ecosystem services that enhance residents' quality of life. These services include air purification, noise reduction, habitat provision for urban biodiversity, and opportunities for recreation and relaxation. Such green spaces, especially those integrated within residential or high-traffic areas, support physical and mental health, offering restorative environments and mitigating the effects of dense urban development (South et al., 2018). Disaggregated green spaces, rather than a single large area, more effectively promote mental well-being in urban environments, endorsing the importance of strategically distributing green spaces within cities to enhance mental health outcomes (Ha et al., 2022). The cumulative presence of these smaller green areas contributes to urban ecological networks, providing critical "*stepping stones*" that facilitate species movement and enhance overall urban resilience (Lynch, 2019).

The current study relies on the hypothesis that SPUGS help regulate temperature in urban areas during heatwaves. To confirm this hypothesis, we aimed at (1) assessing the vertical and horizontal temperature distribution within an urban fabric and (2) determining whether the recorded altitudinal and longitudinal temperature differences can be correlated with the amount of SPUGS surrounding the sampling points.

Data and Methods

Study Area and Local Context

Bucharest is the capital of Romania, with an official population of 2.2 million inhabitants (National Institute of Statistics, n.d.), but with a larger de facto population, considering all the dwellers living in the city. It is located in a plain area that allowed the expansion of built-up areas throughout the decades, with a centralized and systematic approach during the socialist regime (1947–1989) (Gavrilidis, Grădinaru, et al., 2023) and a more sprawling pattern characterizing the last three decades (Kapraras, 2024; Suditu, 2012). The absence of strategic planning since 1990 has led to inconsistent land use, higher residential densities, loss of green spaces, and numerous land-use conflicts (Ianoş et al., 2017; Iojă et al., 2014). Green urban areas in Bucharest represent 5.6% of the urban core, below the European average, with varying per capita availability due to differing definitions of green spaces, and since 1989, green coverage has decreased while urban development and population density have increased (Badiu et al., 2018; Gavrilidis et al., 2022).

In the post-socialist era, Bucharest's urban organization shifted significantly from socialist-era principles. While the socialist city was centrally planned with large-scale residential and industrial projects, compact public spaces, and emphasis on mass housing estates for workers, the post-socialist organization brought a transition to privatization, decentralization, and market-driven urban expansion. This led to urban sprawl, increased commercial developments, and a reduction in public spaces, with less cohesive zoning that accommodated privatized residential and commercial interests, often resulting in land-use conflicts and reduced urban green spaces (Hirt, 2013). Current neighborhood design in Bucharest is a blend of rapid, sometimes chaotic post-communist development, characterized by unplanned high-rise buildings, luxury apartments, and gated communities, especially in affluent areas. This has resulted in significant disparities in quality and accessibility between neighborhoods, with affluent areas in the north contrasted against dense, under-resourced zones in the south, where poverty and inadequate services remain pressing issues (Nae & Turnock, 2011). However, these disparities are narrowing in the early 2020s as, due to land availability, real estate development emerged in the less developed areas of Bucharest.

Post-socialist city development in Bucharest led to the acknowledgment that wide urban green space is harder to be designed as unbuilt areas would be mostly preferred for built-up expansion. Hence, the role of SPUGS became much more relevant in terms of urban landscape mitigation and ecosystem services provision (Gavrilidis, Zakerhaghighi, et al., 2023).

The study area selected for the current work represents an urban fabric within a neighborhood developed during the socialist period—*Drumul Taberei* (Fig. 1). These neighborhoods prioritized both standardized construction and shared green spaces, aiming to create a socialist urban identity distinct from capitalist city layouts, by promoting community and visual order within collective architectural structures. As our focus was on determining SPUGS influence over the urban microclimate, socialist built neighborhoods were the most suited urban fabrics in which such green features would be found. These areas feature dense, mature vegetation that provides higher-quality green spaces compared to other areas in the city (Popa et al., 2022).

UAV Preparation and Sampling

To assess vertical and horizontal variations in air temperature across the case study area, a structured protocol was developed for field activities. The equipment used included a DJI Mini 2 UAV (weighing 250 g) for air temperature measurements at various altitudes, alongside a lightweight (3 g) temperature data logger (iButton) to record air temperature. The temperature data logger was mounted on the UAV with a custom-made structure designed to prevent direct contact between the logger and the UAV, as well as to shield the logger from direct sunlight (Fig. 2).

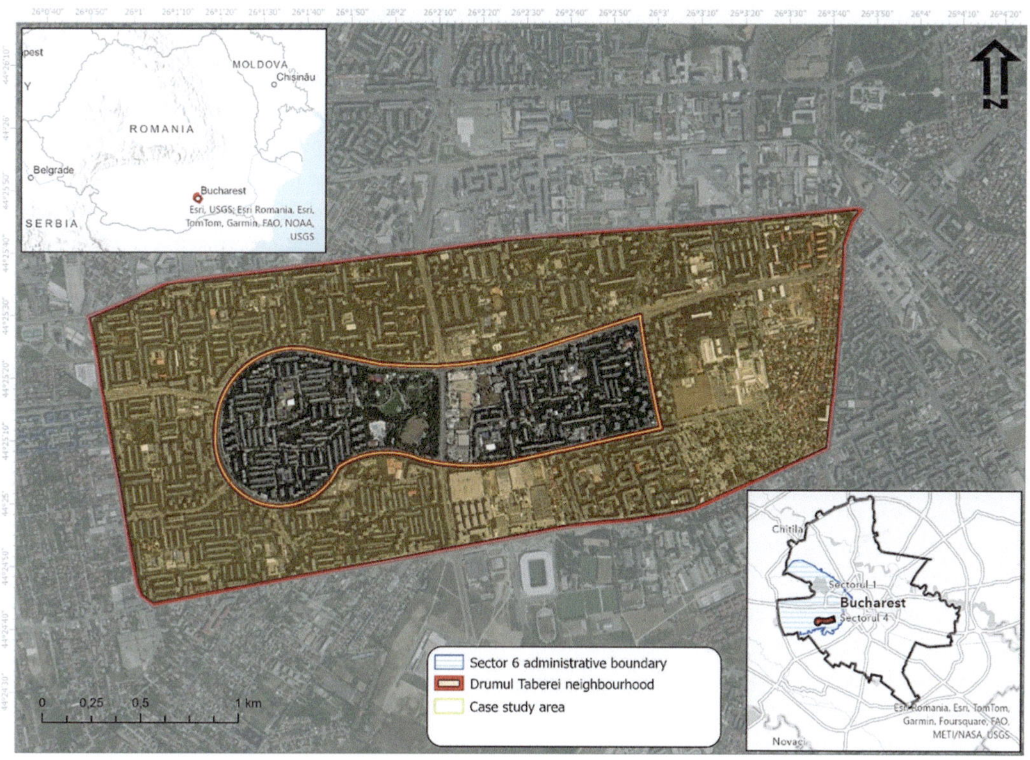

Fig. 1 Case study area position within the city territory

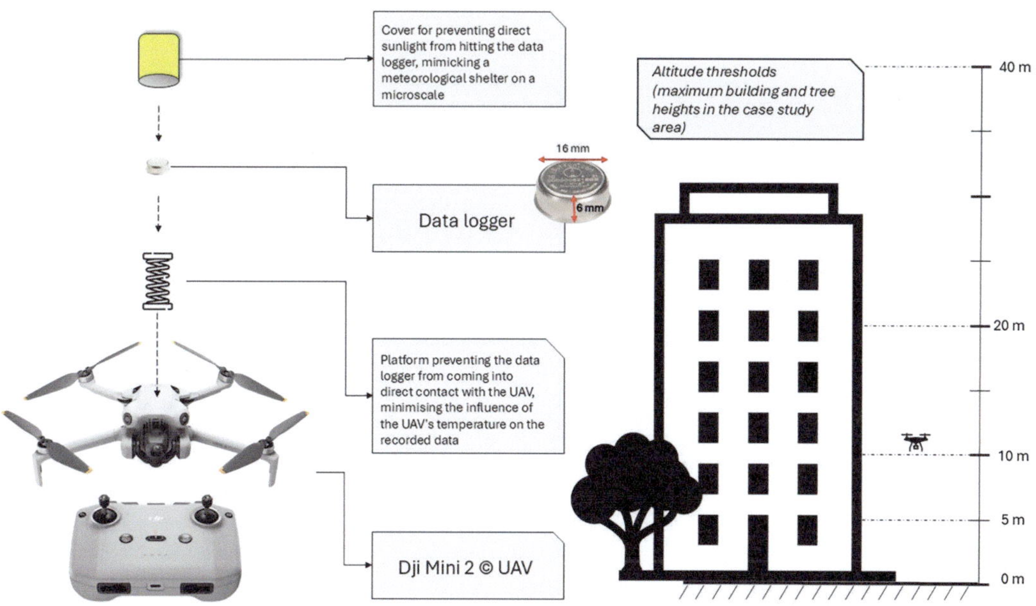

Fig. 2 Customization of the UAV and altitudinal threshold for temperature measurements at each point

This setup aimed to replicate a micro-scale meteorological shelter, enhancing the accuracy of temperature recordings.

Sampling points within the case study area were randomly selected, resulting in a total of 12 points distributed across the study region (Fig. 3). At each sampling point, temperature measurements were taken at five altitude levels: ground level (0 m), 5 m, 10 m, 20 m, and 40 m. These altitude thresholds were determined based on the maximum height of buildings and trees, reflecting the temperature conditions experienced by urban residents at various building floors. The UAV, with the attached data logger, was held at each altitude for 60 s, and the data logger was programmed to record temperature values every 5 s. During measurements at each of the 12 points, a data sheet was completed to record: the official temperature reading for Bucharest at the time of measurement, land use and land cover within a 50 m radius of the sampling point, and the exact time the UAV reached and departed from each altitude threshold. Field activities were conducted on summer days (July–August 2022) when official temperatures exceeded 30 °C.

Data Processing

Following field activities, the recorded temperatures were downloaded from the data logger as a .csv file. The temperature was recorded in Celsius units and the same unit was further used throughout the analysis. Using data from the observation sheets, recorded values were then averaged for each altitude at which the UAV was held for temperature measurements. Subsequently, a spreadsheet was prepared, containing the geographical coordinates of the sampling points, the average temperature recordings for each altitude, the official temperature for Bucharest at the specific time and date recorded by the National Meteorological Administration, and the differences between the logger's data and the official data. In further data processing, differences between the average temperatures recorded by the data logger at each altitude threshold and the official hourly temperature recorded at the Filaret meteorological station (www.meteoromania.ro) were considered. Temperature differences were then spatially interpolated using the IDW method in ArcGIS Pro to provide a spatial perspective. Additionally, a correlation was conducted to

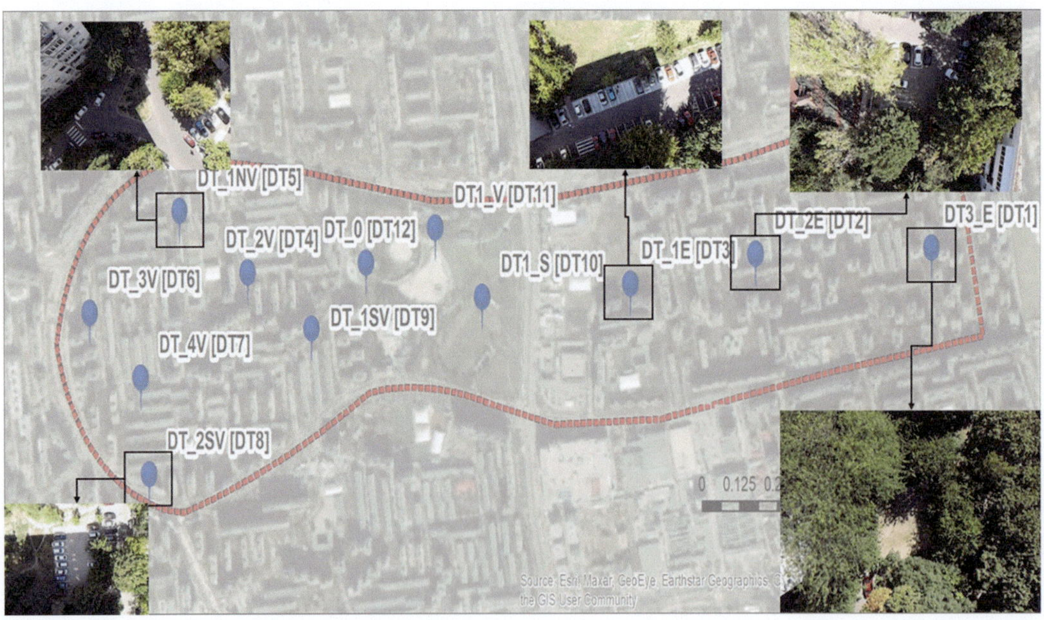

Fig. 3 Sampling points within the case study area

determine whether temperatures at any of the altitudinal thresholds were dependent on the total green or unbuilt area within a 50 m radius around each sampling point.

Results

The main indicator used in our assessment was the temperature values representing the difference between the official recorded temperature and the ones measured by us at each altitude. By that we wanted to determine if a higher share of urban green space surrounding a sampling point would determine relevant differences in temperature. The spatial distribution of the air temperatures differences highlights, the bigger contribution as climate moderator for the park placed in the middle of the urban fabric compared to the areas with SPUGS scattered throughout the neighborhood. Temperature measurements in all five altitudinal layers showed smaller air temperature differences between the recorded and the official data above the park than in other sampling points (Fig. 4). The areas that registered the biggest temperatures differences at every altitudinal layer are those in southwest, near a busy boulevard, and in the east in a high-density urban tissue. The differences between the measured air temperature and the official air temperature in all altitudinal layers are constantly kept somewhere around 6 °C. The mapping of these differences at each altitudinal layer clearly emphasizes how areas with denser concrete infrastructures are subject to higher temperatures of their surrounding air.

For 8 out of 12 sampling points, the average difference between altitude thresholds is negative, indicating that temperature decreases with height (Table 1). For the other four sampling points, the average temperature difference across altitudinal levels shows a positive trend, meaning that from ground level up to 40 m, the temperature records a slight increase. These four sampling points, which record higher temperatures at the 5 and 10 m thresholds, overlap with the only major green area in the analyzed urban fabric—Drumul Taberei Park. Hence, at ground level, temperatures are lower because the tree canopy prevents the ground surface from overheating. However, between 5 and 10 m above ground, in the area just above the tree canopy, the air is warmer. Nevertheless, the share of land covered by vegetation on a 50 m radius is not the highest for the sampling points overlapping the above-mentioned park as there are multiple artificial surfaces (alleys, paved squares, concrete infrastructures) (Fig. 5). It is worth noting that the park has an artificial water body covering 1.1 ha at its center, which modifies the local air humidity levels and could potentially influence air temperatures.

The correlation between the average temperature difference in vertical distribution and the share of land covered with vegetation is −0.713, indicating a strong negative correlation (Fig. 6). This means that as the share of land covered with vegetation increases, the average temperature difference tends to decrease. This finding suggests that areas with a higher percentage of surrounding vegetation tend to experience smaller temperature differences across altitudinal thresholds.

Discussion

The main finding of this study indicates that vertical temperature variation is less pronounced when the land cover below is dominated by vegetation. This aligns with the established understanding that vegetation exerts a cooling effect on the surrounding environment (Silveira et al., 2024). Another significant finding is that the assessment protocol tested in this study produced meaningful results, suggesting that it can be refined and applied at larger scales and in diverse territorial contexts.

Our results show that, at sampling points located within the urban park in our study area, ground temperatures are only slightly higher than the officially recorded temperature, with differences ranging from −1 °C to 2.1 °C. Previous studies have documented that urban parks can reduce daytime temperatures by up to 4.9 °C during warmer periods (Wang et al., 2019), with an

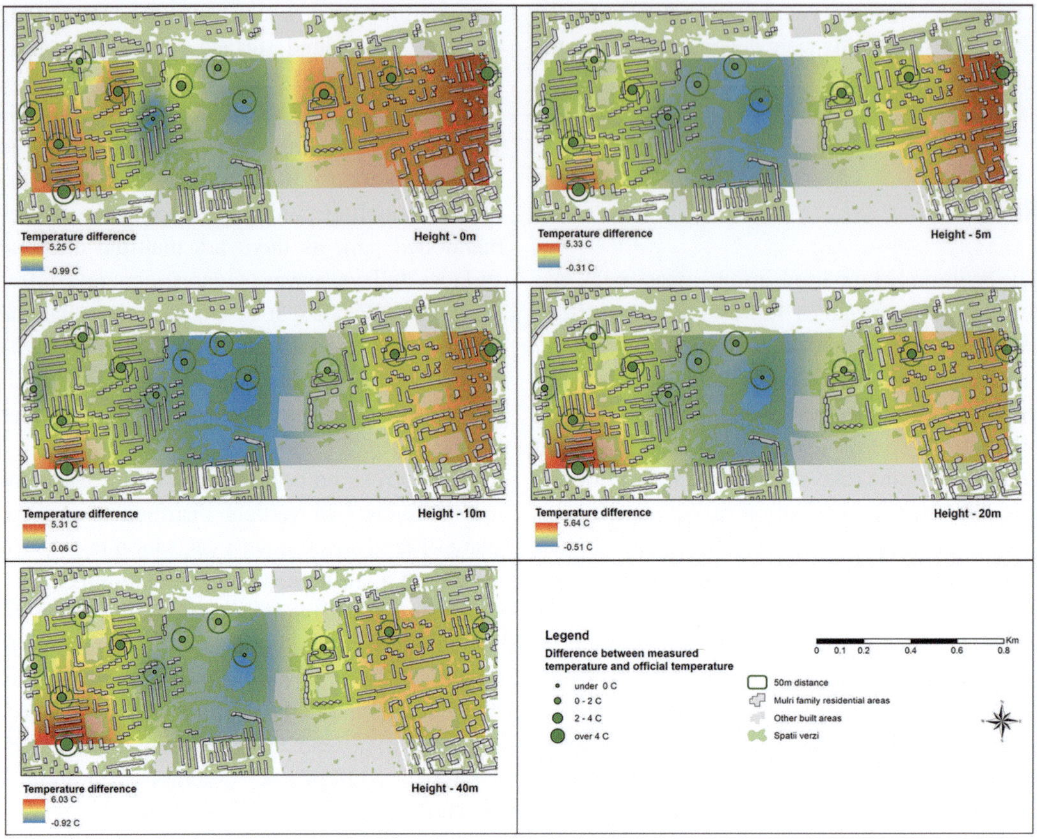

Fig. 4 Interpolation of the difference between the measured values and the official air temperature

average cooling effect of 2.9 °C (Marando et al., 2022). However, the extent to which urban green spaces mitigate elevated temperatures depends on the plant species present, with species exhibiting high transpiration rates and large canopies being particularly effective. Mature trees with expansive crowns are thus optimal for urban cooling (Pace et al., 2021).

The thermal inversion observed at sampling points within the urban park—where air temperatures between 5 and 10 m were significantly warmer than at 0 to 5 m—supports the notion that vegetation reflects sunlight, preventing ground overheating and maintaining cooler air beneath the canopy (Bowler et al., 2010; Georgi & Zafiriadis, 2006). Nevertheless, prior research has shown that urban parks can generate a cold air pool within the first 50 m above ground, resulting in a vertical temperature gradient influenced by the park's green cover (Sugawara et al., 2021). Such differing findings may be attributed to variations in park size and surrounding urban structures, as larger parks are more likely to exhibit a cooling island effect, while smaller parks (under 100 ha) in densely built areas may not.

The presence of a large water body within the park aligns with findings that, like vegetation, water bodies play a crucial role in local cooling (Larondelle et al., 2014). Additionally, as an urban park maintained by local authorities, regular irrigation of turf areas likely contributes to cooling by reducing land surface temperatures (Gobatti et al., 2023).

The observed negative correlation between vegetation cover and temperature variation (−0.713) supports the hypothesis that vegetation reduces local temperature variability (Lin et al.,

Table 1 Average vertical temperature difference per sampling point and the share of green areas in sampling points' proximity

Sampling points	Average temperature difference between altitudinal threshold (°C)	Share of land covered with vegetation in a 50 m radius (%)
DT3_E [DT1]	−0.768	77.05
DT_2E [DT2]	−0.298	73.13
DT_1E [DT3]	−0.480	52.95
DT_2V [DT4]	−0.415	73.33
DT_1NV [DT5]	0.023	54.27
DT_3V [DT6]	−0.503	51.47
DT_4V [DT7]	−0.355	65.63
DT_2SV [DT8]	0.433	31.35
DT_1SV [DT9]	0.193	53.75
DT1_S [DT10]	−0.130	62.50
DT1_V [DT11]	0.003	54.52
DT_0 [DT12]	−0.468	60.92

2016). Vegetated areas typically experience smaller temperature fluctuations, possibly due to shading, transpiration, and reduced heat absorption compared to built-up areas (Egerer et al., 2024). The data suggest that increased vegetation cover may stabilize microclimates, minimizing temperature gradients over short distances. Prior studies have shown that urban canyons with green cover experience less vertical temperature variation, especially in shaded areas.

Vegetation, including street trees, influences vertical temperature profiles by lowering temperatures closer to the ground, with diminishing effects at higher altitudes (Tsoka et al., 2020). This is consistent with our findings, where high-rise residential buildings with small green spaces between them showed limited vertical temperature variation. In our case study area, buildings reach heights of approximately 30 m. Studies have shown that in such dense urban environments, vegetation creates a stronger vertical temperature gradient, with air temperature decreasing more significantly with height than in less vegetated areas, where the urban heat effect reduces vertical temperature differences (Mao et al., 2024).

Our findings contribute to existing literature on the role of small-scale green features in mitigating UHI effects. By optimizing the size, shape, and distribution of these small green areas, cities can enhance cooling effects, particularly in compact neighborhoods (Murtinová et al., 2022). Through shading and evapotranspiration, small green spaces help reduce urban heat and improve residents' thermal comfort at a local scale (Picard & Tran, 2021). Narrow, vegetated alleys provide natural cooling and enhance community attachment, creating cooler, more livable urban environments (Weber & Schneider, 2021).

The neighborhood used in this study is one of the "greenest" in Bucharest and is generally associated with a high quality of life (Onose, 2018). The existing SPUGS are a legacy of socialist-era planning when the area was designated for military personnel and their families. This planning legacy distinguished the neighborhood from other working-class areas in terms of livability. Despite the reduction in SPUGS over the past 30 years—due to the creation of parking lots and wider roads—this heritage continues to add value for residents and potential property buyers (Badiu et al., 2018).

When interconnected, small green patches function as a network of "green heat sinks," effectively mitigating urban heat. Enhancing the connectivity of green spaces allows urban areas to better regulate local temperatures and improves resilience against heat waves (Mokhtari et al., 2022). In our study area, an urban park situated at the neighborhood center is linked by SPUGS that extend to the neighborhood's periphery. Studies on the UHI effect in Bucharest have shown that this urban fabric records lower land surface temperatures than surrounding urban areas (Grigoraș & Urițescu, 2019).

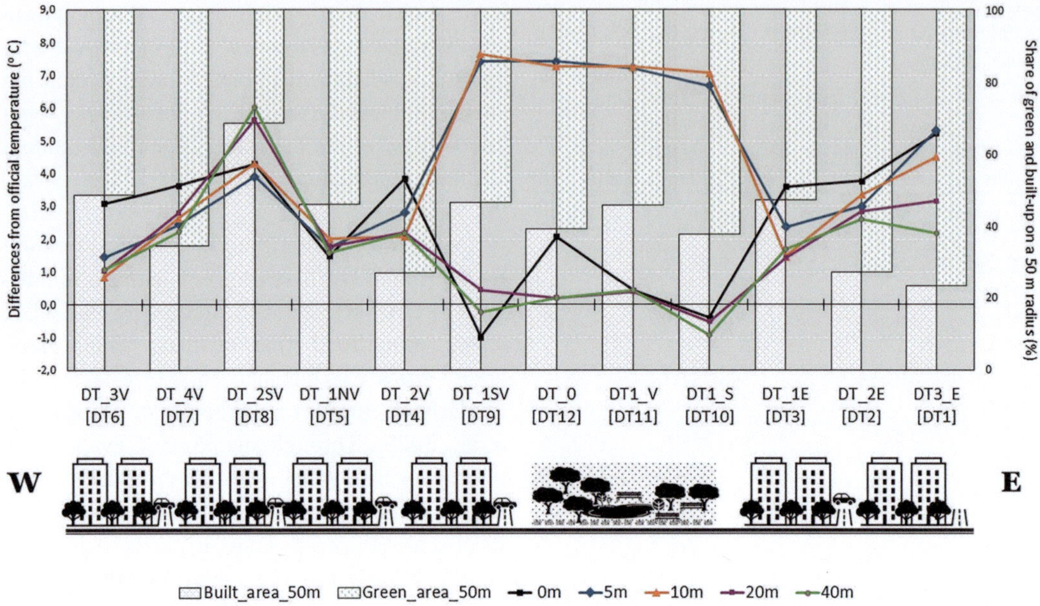

Fig. 5 Temperature difference distribution and land cover shares per each sampling point on a west–east profile of the study area

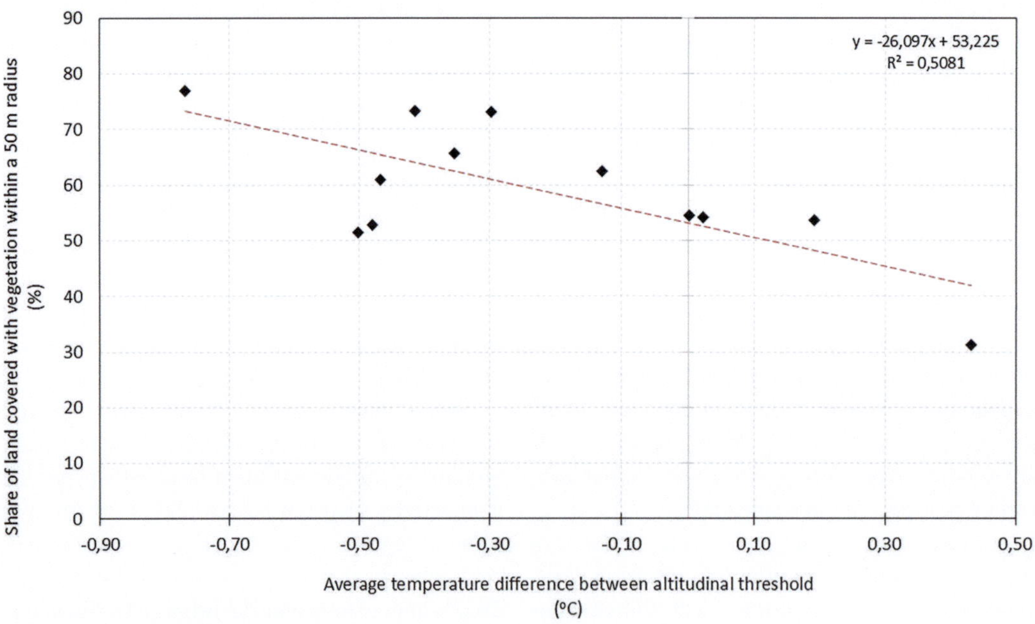

Fig. 6 Relationship between temperature difference and vegetation coverage

Strengths and Limitations

The main strength of our analysis is the establishment of a field protocol for measuring vertical temperature profiles that does not require expensive equipment. The primary costs included the purchase of a DJI Mini 2 UAV, an iButton temperature data logger, and three spare UAV batteries. The total expense for all field equipment amounts to approximately €750–800. In addition to the reduced costs, the use of a UAV weighing under 250 g allows operation with minimal regulatory requirements—registration on the DJI application is sufficient, without the need for additional flight pilot certifications. Furthermore, lighter UAVs face fewer restrictions when flying over densely populated urban areas. This methodology demonstrates feasibility for three-dimensional mapping of air temperature distribution in urban environments.

A limitation of our methodology is that a larger number of sampling points would have increased the robustness of the data and allow for more definitive conclusions. Additionally, by focusing on a single type of urban fabric within a city as large as Bucharest, the generalizability of our findings to the entire city and the effects of SPUGS on urban air temperature regulation is limited. However, a follow-up study with twice as many sampling points and involving different urban fabrics across the city is currently underway, with results pending publication. This initial analysis served as an experimental approach to validate the field methodology and data processing techniques.

Conclusion

The hypothesis proposed in this study has been confirmed by the results obtained: the presence of SPUGS in urban settings impacts air temperature both at ground level and at various altitudes. This data provides valuable insights for urban planners aiming to reduce UHI effects, suggesting that increasing vegetation in urban areas can help stabilize temperatures and minimize fluctuations. Such findings are particularly relevant for dense, compact cities where planning large urban green spaces presents a significant challenge. Additionally, this information could be beneficial for environmental assessments, especially in areas where managing temperature variability is essential for biodiversity conservation.

References

Anwar, M. M., Hashim, M., Aziz, A., Stocco, A., Abdo, H. G., Almohamad, H., Al Dughairi, A. A., & Al Mutiry, M. (2023). Urban green spaces distribution and disparities in congested populated areas: A geographical assessment from Pakistan. *Sustainability, 15*, 8059.

Badiu, D. L., Onose, D. A., Niță, M. R., & Lafortezza, R. (2018). From "red" to green? A look into the evolution of green spaces in a post-socialist city. *Landscape and Urban Planning, 187*, 156–164.

Bai, Y., Wang, W., Liu, M., Xiong, X., & Li, S. (2024). Impact of urban greenspace on the urban thermal environment: A case study of Shenzhen, China. *Sustainable Cities and Society, 112*, 105591.

Bowler, D. E., Buyung-Ali, L., Knight, T. M., & Pullin, A. S. (2010). Urban greening to cool towns and cities: A systematic review of the empirical evidence. *Landscape and Urban Planning, 97*, 147–155.

Egerer, M., Annighöfer, P., Arzberger, S., Burger, S., Hecher, Y., Knill, V., Probst, B., & Suda, M. (2024). Urban oases: The social-ecological importance of small urban green spaces. *Ecosystems and People, 20*, 2315991.

Ettinger, A. K., Bratman, G. N., Carey, M., Hebert, R., Hill, O., Kett, H., Levin, P., Murphy-Williams, M., & Wyse, L. (2024). Street trees provide an opportunity to mitigate urban heat and reduce risk of high heat exposure. *Scientific Reports, 14*, 3266.

Evans, D. L., Falagán, N., Hardman, C., Kourmpetli, S., Liu, L., Mead, B., & Davies, J. (2022). Ecosystem service delivery by urban agriculture and green infrastructure–a systematic review. *Ecosystem Services, 54*, 101405.

Fang, X., Li, J., & Ma, Q. (2023). Integrating green infrastructure, ecosystem services and nature-based solutions for urban sustainability: A comprehensive literature review. *Sustainable Cities and Society, 98*, 104843.

Fotoohi, O., Jahani, N., & Lee, S. (2024). A method for regulating land surface temperature by creating balance in urban landscape heterogeneities: The case of Tehran metropolitan. *Environment, Development and Sustainability*. https://doi.org/10.1007/s10668-024-05334-6

Gavrilidis, A. A., Grădinaru, S. R., Onete, M., Breuste, J., & Ioja, C. (2023). The influence of vacant land presence on proliferation of invasive alien plant species: The case of ailanthus Altissima (mill.) Swingle Var. Altissima. In *Making green cities: Concepts, challenges and practice*. Springer.

Gavrilidis, A. A., Popa, A.-M., Onose, D. A., & Gradinaru, S. R. (2022). Planning small for winning big: Small urban green space distribution patterns in an expanding city. *Urban Forestry & Urban Greening, 78*, 127787.

Gavrilidis, A.-A., Zakerhaghighi, K., Popa, A.-M., Akbarian, S. Z., Onose, D.-A., Grădinaru, S. R., & Slave, R.-A. (2023). Perceptions of cultural ecosystem services provision by small public urban green spaces: Perspectives from different cultural backgrounds. *Urban Ecosystems, 27*, 699–716.

Georgi, N. J., & Zafiriadis, K. (2006). The impact of park trees on microclimate in urban areas. *Urban Ecosystems, 9*, 195–209.

Gobatti, L., Bach, P. M., Scheidegger, A., & Leitão, J. P. (2023). Using satellite imagery to investigate blue-green infrastructure establishment time for urban cooling. *Sustainable Cities and Society, 97*, 104768.

Grigoraş, G., & Urițescu, B. (2019). Land use/land cover changes dynamics and their effects on surface urban heat Island in Bucharest, Romania. *International Journal of Applied Earth Observation and Geoinformation, 80*, 115–126.

Ha, J., Kim, H. J., & With, K. A. (2022). Urban green space alone is not enough: A landscape analysis linking the spatial distribution of urban green space to mental health in the city of Chicago. *Landscape and Urban Planning, 218*, 104309.

Hirt, S. (2013). Whatever happened to the (post) socialist city? *Cities, 32*, S29–S38.

Ianoş, I., Sorensen, A., & Merciu, C. (2017). Incoherence of urban planning policy in Bucharest: Its potential for land use conflict. *Land Use Policy, 60*, 101–112.

Iojă, C. I., Niță, M. R., Vânău, G. O., Onose, D. A., & Gavrilidis, A. A. (2014). Using multi-criteria analysis for the identification of spatial land-use conflicts in the Bucharest metropolitan area. *Ecological Indicators, 42*, 112–121.

Kabisch, N., van den Bosch, M., & Lafortezza, R. (2017). The health benefits of nature-based solutions to urbanization challenges for children and the elderly–a systematic review. *Environmental Research, 159*, 362–373.

Kapraras, G. (2024). *Urban sprawl in Romania: The case of Bucharest* (Vol. 15). Urbanism, Architecture, and Constructions.

Larondelle, N., Hamstead, Z. A., Kremer, P., Haase, D., & McPhearson, T. (2014). Applying a novel urban structure classification to compare the relationships of urban structure and surface temperature in Berlin and New York City. *Applied Geography, 53*, 427–437.

Lin, B. B., Meyers, J., Beaty, R. M., & Barnett, G. B. (2016). Urban green infrastructure impacts on climate regulation services in Sydney, Australia. *Sustainability, 8*, 788.

Liu, W., & Song, Z. (2020). Review of studies on the resilience of urban critical infrastructure networks. *Reliability Engineering & System Safety, 193*, 106617.

Lynch, A. J. (2019). Creating effective urban greenways and stepping-stones: Four critical gaps in habitat connectivity planning research. *Journal of Planning Literature, 34*, 131–155.

Mao, R., Ma, Z., Ning, H., & Cao, J. (2024). Exploring the natural ventilation potential of urban climate for high-rise buildings across different climatic zones. *Journal of Cleaner Production, 475*, 143722.

Marando, F., Heris, M. P., Zulian, G., Udías, A., Mentaschi, L., Chrysoulakis, N., Parastatidis, D., & Maes, J. (2022). Urban heat Island mitigation by green infrastructure in European functional urban areas. *Sustainable Cities and Society, 77*, 103564.

Marando, F., Salvatori, E., Sebastiani, A., Fusaro, L., & Manes, F. (2019). Regulating ecosystem services and green infrastructure: Assessment of urban heat Island effect mitigation in the municipality of Rome, Italy. *Ecological Modelling, 392*, 92–102.

Mokhtari, Z., Barghjelveh, S., Sayahnia, R., Karami, P., Qureshi, S., & Russo, A. (2022). Spatial pattern of the green heat sink using patch-and network-based analysis: Implication for urban temperature alleviation. *Sustainable Cities and Society, 83*, 103964.

Murtinová, V., Gallay, I., & Olah, B. (2022). Mitigating effect of urban green spaces on surface urban heat Island during summer period on an example of a medium size town of Zvolen, Slovakia. *Remote Sensing, 14*, 4492.

Nae, M., & Turnock, D. (2011). The new Bucharest: Two decades of restructuring. *Cities, 28*, 206–219.

National Institute of Statistics. (n.d.). *Tempo online database*. National Institute of Statistics.

Onose, D. A. (2018). *Modelarea interdependețelor dintre componentele structurale, funcțiile urbane și calitatea mediului din municipiul București*. Editura Etnologică.

Pace, R., de Fino, F., Rahman, M. A., Pauleit, S., Nowak, D. J., & Grote, R. (2021). A single tree model to consistently simulate cooling, shading, and pollution uptake of urban trees. *International Journal of Biometeorology, 65*, 277–289.

Pandey, B., & Ghosh, A. (2023). Urban ecosystem services and climate change: A dynamic interplay. *Frontiers in Sustainable Cities, 5*, 1281430.

Picard, P. M., & Tran, T. T. H. (2021). Small urban green areas. *Journal of Environmental Economics and Management, 106*, 102418.

Popa, A. M., Onose, D. A., Sandric, I. C., Dosiadis, E. A., Petropoulos, G. P., Gavrilidis, A. A., & Faka, A. (2022). Using GEOBIA and vegetation indices to assess small urban green areas in two climatic regions. *Remote Sensing, 14*, 4888.

Ribeiro, P. J. G., & Gonçalves, L. A. P. J. (2019). Urban resilience: A conceptual framework. *Sustainable Cities and Society, 50*, 101625.

Shaw, B. J., van Vliet, J., & Verburg, P. H. (2020). The peri-urbanization of Europe: A systematic review of a multifaceted process. *Landscape and Urban Planning, 196*, 103733.

Silveira, C., Dias, A. T. C., Amaral, F. G., de Gois, G., & Pistón, N. (2024). The importance of private gardens and their spatial composition and configuration to urban heat Island mitigation. *Sustainable Cities and Society, 112*, 105589.

South, E. C., Hohl, B. C., Kondo, M. C., MacDonald, J. M., & Branas, C. C. (2018). Effect of greening vacant land on mental health of community-dwelling adults: A cluster randomized trial. *JAMA Network Open, 1*, e180298.

Suditu, B. (2012). Urban sprawl - The legal context and territorial practices in Romania. *Human Geographies, 6*, 73–77.

Sugawara, H., Narita, K.-I., & Mikami, T. (2021). Vertical structure of the cool Island in a large urban park. *Urban Climate, 35*, 100744.

Tan, P. Y., & Samsudin, R. (2017). Effects of spatial scale on assessment of spatial equity of urban park provision. *Landscape and Urban Planning, 158*, 139–154.

Tsoka, S., Tsikaloudaki, K., Theodosiou, T., & Bikas, D. (2020). Assessing the effect of the urban morphology on the ambient air temperature of urban street canyons under different meteorological conditions. Application in residential areas of Thessaloniki, Greece. *IOP Conference Series: Earth and Environmental Science, 410*, 012005.

Ulpiani, G. (2021). On the linkage between urban heat Island and urban pollution Island: Three-decade literature review towards a conceptual framework. *Science of the Total Environment, 751*, 141727.

Venter, Z. S., Hassani, A., Stange, E., Schneider, P., & Castell, N. (2024). Reassessing the role of urban green space in air pollution control. *Proceedings of the National Academy of Sciences, 121*, e2306200121.

Wang, Y., Ni, Z., Chen, S., & Xia, B. (2019). Microclimate regulation and energy saving potential from different urban green infrastructures in a subtropical city. *Journal of Cleaner Production, 226*, 913–927.

Weber, E., & Schneider, I. E. (2021). Blooming alleys for better health: Exploring impacts of small-scale greenspaces on neighborhood wellbeing. *Urban Forestry & Urban Greening, 57*, 126950.

Zhang, F., & Qian, H. (2024). A comprehensive review of the environmental benefits of urban green spaces. *Environmental Research, 252*, 118837.

Climate Change, the United Nations' Sustainable Development Goals, and Equity: A Review of Urban Perspective

Asad Aziz and Muhammad Mushahid Anwar

Abstract

Climate change is considered an influential threat in this century of urbanization to the world's population in every society, potentially affecting every individual in diverse groups of urban inhabitants in the coming decades. Through this study, we raise a question to access how climate change amplifies urban inequities in the context of sustainable development goals, which brings the question of urban equities to the forefront. The study finds the rate of population change, growth rate, urban migration, landscape modification, urban infrastructure, even adaptation, and mitigation affect a more specific group of people rather than others. Considering the sensitivity issue, this review case study only considers gender and socio-economic status and their association with related UN goals. The study finds and provides the results concerning process-oriented, social investigation, and current literature on the given topic. Finally, to avoid equity and justice concerns in reaction to the impacts of climate change and SDGs, it is considered a social concern concerning adaptation and mitigations.

A. Aziz (✉) · M. M. Anwar
Department of Geography, University of Gujrat, Gujrat, Punjab, Pakistan
e-mail: asadaziz@uog.edu.pk; mushahid.anwar@uog.edu.pk

Keywords

Climate · Climate change · Development · Equity · Goals · Mitigation · Sustainable development goals · Socio-economic · Sustainable

Introduction

Urban areas are considered melting spots of a diverse group of people with diverse backgrounds, with cultural, ethnic, social economic, and different financial positions. They make a diverse network in a central place that is called a city. These networks are different from one another concerning intra-city diversities, locations, climate, geographical boundaries, social structure, economic and political functions, and local to national development stages (Covich, 2013). In other words, that is called diversity. At a place, all of these networks are interconnected with each other through different means due to which climate change gets importance and opens the doors for discussion of how these cities and urban diversities interrelate with each other to react climate change and challenge to achieve the sustainable development goals (Tuckett, 2009). This question demands a systematic way of investigation including "who will be the most susceptible group due to change of climate" and the second is "how to address these inequalities

M. M. Anwar, A. Aziz (eds.), *Sustainable Landscape for Urban Resilience*, Sustainable Development Goals Series, https://doi.org/10.1007/978-3-031-99836-2_4

and injustice in these urban diversities" for goals 3 and 5 (Abbass et al., 2022; Wu et al., 2018). For better understanding, this study divides the investigations into two diverse categories that include cities, and then discusses respectively. For who: here, the urban areas defined as high-income countries contributed the most to climate change and low income in the south get impacted the most relating to goal 10 (reducing inequalities) (Wan et al., 2022). Further to this, for climate change, the study considered natural hazards like flooding, droughts, heat waves, and landslides, which have an impact on urban areas differently depending upon their individual, social, and economic functions and factors.

The New Urban Agenda focuses on sustainability in the future through well-planned and managed urbanization in developed and developing countries, aiming to achieve the United Nations' sustainable development goals (Nation, 2017). The most pressing issues, such as rapid urbanization and its impacts on climate change and police for mitigations, cities, communities, and economy, must be controlled, and there should be equity for all. This equity is the fundamental key to sustainable development and is necessary to make cities the heart of sustainable development (Nation, 2022). The major focus of this equity is the provision of all environmental services and protection of natural hazards regardless of income, race, ethnicity, and other such characteristics (Callahan et al., 2021). This focus has three actions at global, local, and individual scales to accelerate and mobilize sustainable solutions to minimize the challenges in the world (Mohammad, 2022). The climatic change mitigation or climate justice associated with environmental equality is still disputed. Low-income communities are predicted to be disproportionately impacted by climate change compared to those who contribute more to greenhouse gas emissions (Adnan et al., 2024; Ullah & Lin, 2024). However, if conclusive actions are taken in this regard, these same communities could gain proportionally from addressing and adapting to climate change.

For how: The word "how" is related to the change of climate and its impacts on urbanities, their risks when they affect, and also their measure of mitigation and adoption (Khan et al., 2024). This impact is more focused on the urban world, where the population of different abilities, capacities, and resources is concentrated (Gupta, 2016). Mitigation and adoption, if not properly defined or designed, will extremely affect vulnerable populations, which therefore manifest inequalities and inequities in urban areas (Nuru et al., 2021b). On the other hand, if the policies are well defined, this will decrease equity concerns of climate change and alleviate the burdens of addressing the most susceptible urban populations to achieve Sustainable Development Goal 10.

The UN goal 1 associated with ending poverty and climate change is linked with each other in diverse means (Leal Filho et al., 2021). This may include the loss of property in response to natural disasters, death, hunger, famine, and diseases that create challenges for poor urban in the world. This situation creates regional injustice, particularly in the developing world, where the national government is unable to meet all of these damages. Reconsidering SDG-3 to ensure health and promote well-being for all ages across the globe, climate change strategies are needed to be adjectives at a local level in urban slums where migration patterns are increasing (Hák et al., 2016; Morton et al., 2019). It has been noted that older people are affected more than younger people due to malnutrition and lack of medical facilities (Rocha et al., 2022).

Addressing the issues of equality and equity has reached global importance, not only acknowledged by the United Nations (Kuh, 2018), but even part of the agreement of post-Kyoto, also known as the Paris Agreement, to accomplish Sustainable Development Goals (SDGs 5 and 10) (Nations, 2016). Extensive progress in the Millennium Development Goals (MDGs) is not only beneficial to uniformly joint or the universal (Nations, 2018). According to the New Urban Agenda, the provision of housing, finance, actions, and planning regarding the climate and provision of basic services is the responsibility of local and regional government (Addo-Atuah et al., 2020). But this scenario is still not acknowledged by NAU (Development, 2016). However, it is found in the literature that gender inequality and extreme poverty persist (Eastin, 2018). Moreover, wide-ranging gaps in all world regions in women's access to paid jobs are quite persistent issues (Eneji et al., 2021). While in some developing regions poor men also get the effect of the impacts of climate change in the form of loss of property, jobs, and economy while facing the disastrous climate change (Wernberg et al., 2016).

This study examines the equality and equity of climate change in the frame of urban areas. For this examination, the study focuses on policies on adoption and mitigations concerning equity in urban areas with a focus on poverty and gender (Nuru et al., 2021a). The fundamental attempt of this study is to provide the crucial background of equity concern in urban areas by disentangling multiple equity perspectives and their association with UN goals. This equity concerns gender, socio-economic status, developing and developed nations, regarding their climate change strategies, and city resilience (Adom, 2024; Kawamoto & Kanie, 2020). Finally, the study symmetrically summarizes diverse influences of climate change, climate hazard, and their differential adaptation policies with outcomes of mitigation among various groups of communities on the urban population. Afterward, the study summarizes and finalizes to highlight the policy implications for addressing cities equitably concerning climate change.

Background

This section explains the domains, different major types, and principles related to equity as defined in the previous research literature and research work. The following are the major factors explained related to the case study.

Equity Domains, Principles, and Types

For numerous local to regional climate initiatives, for current and forthcoming generations endorsing equity is implicit (and sometimes explicit) (Nation, 2015). However, it is still uncertain which type of equity is referring to, which is the reason why this equity is further divided into three more subtypes. Table 1 further elaborates the main focus, costs, and benefits of equity realms.

1. Consequential/distributive outcome-based equity: This equity is related to an action or developmental trend and consequences of a policy (Nation, 2015). This type of equity is related to the distribution of benefits and costs, and burdens between men and women, between households, generations of urban residents, and urban boundaries like urban and peri-urban districts (Huynh et al., 2021).
2. Procedural and process-oriented equity: This type of equity is related to justice and administration related to the decision-making process (Metz, 2000).
3. The third type of this equity is contextual equity that considers the social, economic, and preexisting political conditions into account.

To operationalize the concern of these equities, "Mahanty, McDermott, and Schreckenberg"

Table 1 Generally functional equity realms in international climate change

Focus	Costs (responsibilities)	Benefits
Origins of the main issue	1. Ethical obligation ("guilt" for having or initiated the problem	3. Prior influences to provide the benefits to local community
Significances of the resolution	2. Competencies (capability to *donate* for issues resolving)	4. Essential for the outcomes to be accomplished

Source: Kallbekken et al. (2014), *Equity and Spectrum of Mitigation Commitments in the 2015 Agreement,* TemaNord, Copenhagen, 69 pages

linked the above parameters into three types, A. The equity goal is B.

The measure is related to equity and all the procedures to set the goals. This operationalization is based upon some more principles and indicators (Klinsky & Dowlatabadi, 2009). These indicators and principles are different for every country around the world (Kallbekken et al., 2014), in different geographical scales, including local to regional, from urban to rural, and even from gender as shown in Table 2 to underline the principle of contextual equity, that is, goal C. These underline the principles such as the "ability to pay and the exemption of the poorest."

The inclusion of related stakeholders, concerning the need for consideration related to process-associated equity such as women's participation in bodies, boards, and gender balance, is still unequal even in principles of gender balance (Eneji et al., 2021). However, in "UNFCCC" the groups' women's rights have made significant contributions to several decisions for participants of women in their adaptation measures, mitigation plans, technology, capacity building, and climate change finance to decide their thematic plans. Anyhow, determinations concerning adaptation take rebut gender-sensitive language. However, the mitigation relation to gender still needs further gender-sensitive mitigation actions (Organization, 2014).

Environmental justice, equality, and equity are the major social issues that create a debate on climate change, and this debate has recognized that the countries that contribute least to climate change might be impacted by the climatic issue in the future or near future (Anser et al., 2023; Satterthwaite et al., 2014). The strategy "Climate Change Resilience" has been developed in recent years to desire to minimize the impacts of climate

Table 2 Clarification of vital gender perceptions

- Sex is a genetic notation that refers to the biological variances in men and women in terms of reproductive and sexual functions
- Gender: Gender discrimination states to the generally synthetic distinctions in men and women. Gender traits, characters and connections, and power dynamics are all part of it. Gender roles fluctuate significantly between cultures and communities, shaping the gender distribution of employment access, control usage, and ownership of resources, opportunities, and products. All of these variables are used to examine the social developments and changes across time
- Gender equality means the men and women have identical chances to realize their full human rights, as well as that society values both the resemblances and differences in men and women, and the various parts that they perform
- Gender equality is the method of treating men and women equally. To maintain justice, it is frequently important to put procedures in place that recompense for societal and historical difficulties that hinder women and men from competing in all field
- Gender sensitive means to examine socio-cultural standards and judgment to identify varied moralities, duties, and obligations of men and women in the community, as well as the interactions among them. Gender-sensitive strategies, projects, financial and administrative actions, and structural practices will differentiate in men's and women's capacities, requirements, and priorities; ensure that both men's and women's views and ideas are occupied completely; reflect the inferences of results on the condition of women comparative to men; and take arrangements to state disparities or imbalances among genders
- Gender mainstreaming means the approach to evaluate the consequences of several strategic for men and women, at all levels, and as an approach for attempting to make men's and women's concerns and opinions an vital dimension in decision-making, monitoring, implementation, and evaluation in financial, political and social circles, so that men and women benefit similarly and disparity is not propagated. Gender equality is the critical goal. Gender mainstreaming was first presented in the fourth UN world conference on women in 1995 in Beijing. It was incorporated into all United Nations policies and programmers in 1997

Sources: GGCA (2009), Training Manual on Gender and Climate Change, International Union for Conservation of Nature (IUCN), United Nations Development Programme (UNDP), and the Global Gender and Climate Alliance (GGCA), 262 pages; also WEDO and REDD+SES (2013), From Research to Action, Leaf by Leaf: Getting Gender Right in the REDD+ Social and Environmental Standards (Booklets 1 and 2), Women's Environment & Development Organization (WEDO), New York

change on the community at risk (Al-Humaiqani & Al-Ghamdi, 2022; Baraj et al., 2024). This can be associated with the poor communities in equity concern in such a way that the poor people get worse from floods, droughts, and sea level rise as they have less ability to resist the changes (Fahy, 2021). In other words, this equity or environmental injustice is also linked with "Climatic Gentrification." This debate proved that the low-income people get affected the most and climatic impacts are also increased in these countries because of supply chain interdependencies and the absence of environmental laws (Nabangchang et al., 2015). Initially, this discussion is resolved around the mitigation responsibility. Most scientists (Metz, 2000) put great stress that climate change only considers adaptation and impacts but also mitigations. Differential impacts and adaptation need particular importance at a different scale from local to regional geography. Adaptation, mitigation, impacts, and risks are the four major domains that get consideration for equity in the current urban climate change debate.

Methods

The methodology of the current review study is based on the recent scientific literature on adaptation, mitigations, and climate change, particularly in cities and their association with mitigations, adaptation plane, environmental justice, and equity. This portion is the conceptualization of the topic and provides the base for further proceeding. Adoption and mitigation are diverse approaches that have multidisciplinary concepts in different studies published in the past. The study integrates these approaches associated with UN goals and equity for an urban area from the reputed database such as Web of Science, Scopus, UN reports/documents, including UN Agenda 2030, New Urban Agenda (NUA), World Urban Forum Declarations, Decade of Actions, and other government documents. The facts used in this study consist of scientific publications during the past few years, in peer-review top-ranking journals, including *Science Direct/Elsevier, Springer, Emerald, MDPI*, and some other quality research publishers, controlled by the scientist. This data includes scientific publications, official reports, review papers, and short communication, but also some gray literature, including conferencing and working papers, and even also the conferencing proceeding. On the way to create an equilibrium in the literature review process to cover the whole world, the study considered the papers from low-, middle-, and high-income states, and related to the topic of the study. To refine the whole database, the duplication of the data was checked and duplicate papers/concepts were removed to create harmony in the data base. After refining, the validity of these papers was checked by reading the titles, abstract, and introduction of the papers. Refining provides the exact data associated with the proposed study objectives. The study followed the modeling and methods used in the past study and also highlighted the gaps in past literature. These gaps include the impacts of climate change on men's, environmental injustice, and developing countries with the multidimensional socio-economic profile associated with participation in climate change actions and UN goal 17. After a deep analysis, the modeling of outcomes is presented in the findings.

The major review portion of this study comprises a valuation change of climate and its consequences on urban areas and also analyzes in what way it resists achieving UN sustainability goals in international exercise. This assessment was associated with the proposed objectives of the study, including good health and well-being (goal 3), gender equality (goal 5), reducing equality, sustainable housing (goal 11), climate change and actions (Goal 13), and participation across the globe for climate action (goal 17). To achieve the desired objectives, the study focused on adaptation, climate change impacts, gender equality, poverty, and mitigation policies, as well as the socio-economic profile distribution in built-up areas. Concerning our focused factors as mentioned above, the full details and findings are mentioned below.

Outcomes of the Research Study

On the nature of outcomes, the study divides the results into more classes such as the following:

Across the urban areas in the world, the effects and risks of change of climate like dangerous weather conditions and their effects gradually change (Fuchs, 2010) due to the following factors:

1. The location of a community is determined by physical exposure.
2. The characteristics of populations in an urban area are related to demographic, economic, and social factors (Barrios et al., 2006).
3. The institutional, governance, and power aspects (Kern & Bulkeley, 2009).
4. The construct risk and process involved in urban development (Eiser et al., 2012).

The above factors are associated with playout in small and large urban areas, as well as in low-middle and high-income states (Adelekan, 2019).

The study found that there is evidence that risky weather conditions and climate change disproportionately affect the population, especially women in low-income countries with a low social status (Reckien et al., 2014). Evidence proved that in the low-income countries, including Alexandria, Cairo, Dhaka, and Rio de Janeiro, the population with low social status inhabit areas, exposed to more climatic risk (Bartlett et al., 2009). The risk of exposure is closely related to poor-quality buildings, high population densities, absence of services, previous disaster failure to draw or implement lessons, and lack of risk-reducing infrastructure (David & Enarson, 2012). The poor people in slums are more at risk from natural hazards and high risk of climate change. The low socio-economic status and bad government infrastructure in vulnerable lands in cities make the urban world more at risk and create challenges for urban resilience (Baker, 2012). Together, management shortfalls with bad governance result from an accumulation of risk over time (Singh et al., 2010). This factor is well documented by the previous record of increasing disaster losses due to earthquakes (Reckien et al., 2017), floods, mega-debris flows (Reduction, 2009), tropical storms, and flash flooding in cities (Revi, Satterthwaite, Aragón-Durand, Corfee-Morlot, Kiunsi, Pelling, & Solecki, 2014). However, natural events occurred regularly like floods (Rao et al., 2013), droughts, and flash flooding steadily demoralized the resource foundations of rich groups (Tompkins et al., 2013).

With other social susceptibility markers, gender and poverty censoriously overlap with others. Low-income women are comparatively more vulnerable than high-income women (Abbasi et al., 2019). Men are less susceptible than women, and the results found a negative association with SDG-5, related to gender equality, particularly in the developing world. Similarly, women living in informal settlements are more vulnerable than women living in a formal setting. Adult healthy women have a less risk or exposure than disabled children or men (Goodrich et al., 2019).

Sustainable Development Goal-3 and Equity Concerns

One of the major hazards in the urban world associated with climate change is rising temperature, which covers two main dynamics: one is urban heat island, which means the variation in the rural and urban areas and an average increase in global temperature. According to the directions of "The Glasgow Climate Change Conference" (GCOP-27), all the nations in the world should take collective efforts to limit the global rise in temperature (Change, 2022). These dynamics can provide benefits if related to reducing morbidity and mortality risks of cold temperature (Li et al., 2020; White-Newsome et al., 2011); they also result in heightened mortality and morbidity during extreme heat or heat wave periods. Heat waves in urban areas create additional fatalities and major climatic risks by which hurricanes and floods occur as a result of their consequences (Campbell et al., 2018).

Heat waves in cities cause two types of impacts, including direct effects such as heat stress and indirect effects such as increased mortality and morbidity (Campbell et al., 2018). This direct heat stress is very harmful particularly for humans at night time when high temperatures prevent the human body from regeneration, sleeping, and responding (Amengual et al., 2012). Through the interaction of heat and associated environmental issues, indirect effects on health, particularly air and water pollution, arise principally (Rankoana, 2020). Good health and well-being achieving the UN SDG 3 give stress to managing the hazardous material and chemicals in urban settings, minimizing environmental pollution, and creating sustainability. The urbanization rate in the developing world is unpredicated and unsequenced due to faster population growth. This rate of population is a challenge for well-being and prosperity in the urban areas for all genders.

Across the population, the heat-associated risk is stratified and associated with "intrinsic" and "extrinsic" factors (Fernandez Milan & Creutzig, 2015). These stratified factors need to be minimized to achieve SDG 10 to reduce inequality, particularly in the developing world. This inequality needs cross-border, cross-political, social, and economic corporation among the states for development and equality.

This mutual corporation without gender, religion, color, and social discrimination is needed to combat the actions or activities that alter the climate and climate change to achieve the United Nations goal 13. In this case, preexisting medical facilities, female sex, and age were identified as the main factors (Appiah et al., 2022). All of these factors are identified in "18" different studies (Schuster et al., n.d.). A study found that physical fitness is a major factor in determining age. Regarding the woman's sex due to the influence of thermoregulatory differences and physiological impacts, females are more heat intolerant than males. Typically, a woman also has more exposure to the heat as they spend more time in interior spaces where there are no air conditions or adequate airflow for cooking in informal settlements. People who are confined to bed and less mobile are more vulnerable to heat waves in case of their medical status (Zhang et al., 2020). The same is the case of people who are at high risk of cardiovascular diseases (Jabeen & Jabeen, 2018). In the case of extrinsic factors, lower socio-economic status, education, family members, unemployment, overcrowding, occupation, household ownership, as well as lower education level also create exposure to heat stress (Laaidi et al., 2012). The urban heat also excessively influences socio-economically deprived families because their houses where urban green spaces and green infrastructure are far away and they do have not the ability to fund, develop, and maintain private green space (Nitschke et al., 2017). The risk-reducing environments, such as water and open spaces, make their environment and atmospheres cool. Finally, people in the core of the city are more at risk than those living out or in subareas.

Participation, Equity, and Sustainable Development Goal 13 (Climate Action)

Over short timeframes, heavy precipitation can lead to hazards such as landslides and inland flooding, reflecting the impacts of climate change (Dutta & Dutta, 2021). On the other hand, a drought occurred due to a lack of precipitation. On a massive scale, inland flooding occurred in Australia and Thailand in 2011 (Sun et al., 2021), and also in Pakistan in 2010 (Rahman & Khan, 2013). In 2012, in Uganda, substantial damage and threat were caused by localized flash floods to livelihood (Pérez-Molina et al., 2017), health, and lives in urban population (Komori et al., 2012). Informal settlements get frequent flooding as they were developed on vertical slopes and landslides, in many cities around the world (Carcellar et al., 2011). In the case of urban areas of low- and even middle-income states, the inland flood occurred and the features it caused are impermeable surfaces. This action also leads to soil runoff and damage green spaces when absorbing too much water flows as the scarcity of parks and poor drainage system caused by development houses without planning on marshland and the other number of natural buffers in cities (Revi, Satterthwaite, Aragón-Durand, Corfee-Morlot, Kiunsi, Pelling, & Sverdlik, 2014).

In the urban area, poor people are affected more as they live in environmentally riskier areas and their neighborhoods lack risk-reducing measures a major resistance to SDG-10. The associated risk due to living in the canal and urban rivers has more exposure to flood, particularly the poor urban who lack security and where there is the fast-growing urban areas and population (Bonaiuto et al., 2015). In this case, poor countries in Asia get more attention where 505 million people, or 29.7% estimated, live in informal settlements or poor houses found near the banks of rivers or canals (Kousky et al., 2020). Similarly, in the case of Africa, this percentage is higher than in Asia, which is 37.5% and 61.7% in sub-Saharan Africa, respectively (UN-Habitat, 2010).

Proximity to the coastal areas and water bodies also creates another type of social injustice, displaying the poor urban from their livelihoods and homes. Similarly, another indirect effect related to unsanitary conditions is a health risk, causing waterborne diseases like the respiratory problem, skin diseases, and other types of waterborne diseases when hazardous materials get mixed with floodwaters and spill into open walls (Akanda & Hossain, 2012). In dense low-income neighborhoods, acute respiratory infections, dysentery, outbreaks of cholera, and diarrheal diseases have been recorded due to heavy rainfall (Dankelman & Jansen, 2010). It will result in an increase in the number of unpaid hours of women and the period of care work when they spend their time with children and the elderly to care for them and create challenges to achieve UN goal 3 (Tovar-Restrepo & Irazábal, 2013). To cope with these health emergencies and sanitary, many women on many occasions left their jobs.

The triggering landslides that are associated with excessive rainfall are also another risk factor but are not limited to earthquakes, extreme precipitation, and floods (Bürgi et al., 2004). In many countries, major cities create significant environmental threats to human beings. This underestimation adds to the falling awareness among the general public and authorities of their risk. Usually, the combination of locational and geo-hydrological features, and rainfall triggered the landslides, mostly in low-income countries, where maintenance structure is weak. Men and women experience different ways of migration and displacement, for example, women migrate to urban areas when excessive rainfall damages their houses and increases poverty. While in the case of men, they move to cities to make more money. Finally, delayed or insufficient precipitation harshly impacts the local populations through water shortage, which leads to generating crop failures that indirectly increase food price and also affect the low-income population potentially. There is a dire need to take urgent action to minimize the impacts of climate hazards and action for climatic justice to achieve SDG 13.

Social Injustice Leading to Climate Gentrification

The most vulnerable threat associated with climate change and social justice is climatic gentrification, a term used for social injustice, particularly in coastal areas (Kuh, 2018). Around the world, climate change causes the rise in sea level, storm surge in coastal areas, and heavy rain, which create lifetime threats in coastal areas (Eastin, 2018). The major cities in the world, including London, Hong Kong, Mumbai, and Miami, are the most vulnerable (Eiser et al., 2012). It has been noted the poor population was affected mostly when they were displaced from the high-elevated areas due to heavy rent or land values. Informal settlements are frequently impacted rigorously because of poor infrastructure. Coastal flooding is caused by heavy rainfall, causing flooding in river deltas.

Moreover, around the world, increase in air pollution and global warming also change the rainfall pattern around the globe, and sea-level rise due to the increase in temperature and melting of glaciers. Similarly, high waves due to storm surges affect the coastal population very badly.

Compared to rural areas, the urban population is more exposed to sea-level rise as urban areas are mostly at seas or coastal areas. Approximately 18% of the population in Asia live near the coastal zones (McGranahan et al., 2007), and 12% of them are exposed to threats in the near future (Kovats & Akhtar, 2008). Similarly, some of the major cities in Africa and Latin are also exposed to be vulnerable due to climatic change (Güneralp et al., 2015).

Adaptation for Environmental Justice in Urban Areas

Adaptation to climate change is related to environmental justice and equity, and this equity is included in inadequate adaptation, failure to adapt or no adaptation, and maladaptation in urban areas to climate change (Dodman & Mitlin, 2013). These adaptations are associated with environmental justice, the nature of risk, and differences in regional scale. Settlements characterized by access to infrastructure, such as drainage, piped water, sanitation, roads, health facilities, emergency services, education, markets, and affordable housing for low-income groups (Krishna et al., 2014). It is significant for mitigation to recognize how inhabitants in low- and even middle-income countries adopted a change in the climate. It has been noted that in most developing countries where governments and populations have limited capacity and low budgets with less investment capacity it leads to deficiencies in effective urban planning (Krishna et al., 2014); this is elaborated more in Table 3. The lack of risk-controlling infrastructure is also another face of this problem. At the other end of the spectrum, the countries having provisions of all of these facilities with active climate change policies have less exposure to risk but they cover a very small portion in an urban world. However, all of these major cities of a developed nation have better infrastructure, services, and financial systems in the world to increase the level of resilience to boost goal 13 (Qiang, 2014).

In the end, it is also not true that all discriminations in threat are addressed in ecological aspects even in developed countries. In many cities of low- to middle-income countries, the risk factors also show their impacts due to inadequate infrastructure (Bulkeley et al., 2014). These include poor management and policies, lack of awareness, location of houses in informal settlements, and dangerous locations like flood plains, and coastal areas (Lawhon, 2014). In these areas, the factors involving equity, unclear land tenure, and property rights in urban areas like Dakar, Dar es Salaam, Nairobi, Manila, and also in Kolkata have been documented (Schlosberg & Collins, 2014). It is not only the government but also the city or national governments' choice not to underpin attention to climate change adaptation (Ajibade et al., 2013).

Within cities, risk differentials are also related to health status, sex, gender, and age (Nations,

Table 3 In what way climate change relates to cities' poverty eradication/reduction

Equity concerns for poverty eradication	Solutions
Concerns about equity originate in climate risks and consequences in urban areas, as well as adaptation and mitigation methods that ignore their implications on the urban poor, such as: • Heat: Usually affect the old and women • Floods: Frequently concerning to low income and relegated populations (in LIC and MIC) through high contact and low surviving • Landslides: impacting several revenue groups (MIC, LIC, and MIC) through high contact and low defense • Storm surges: disturbing poor even disregarded groups • (HIC, MIC, and LIC) through high contact and low defense, but similarly wealthy families in HIC. Inside urban areas, low-income neighborhoods at more risk areas	Adaptation: Make the primary objective of adaptation to safeguard the most vulnerable urban communities Provide enough structure and services in all urban areas and effort against authorities' thoughtful failure to do so Align adaptation with proper financial, institutional, and economic support and incentives Mitigation: Stop a disproportionate influence on the urban poor, for example, via changes in housing affordability, prices, and transit access Compensation should be provided in low-income towns, urban areas, and regions (across the region and within countries)

Note: LIC, MIC, and HIC: low-income, middle-income, and high-income countries respectively
Adapted from Reckien et al. (2017)

2018). For example, in the case of sex, the women in the floods of Lagos in 2011 are more vulnerable than men. The people living in the informal settlement lack attention of the government. Thus, adaptation measures are acknowledged among the experts and contributed to reducing the risk measurements.

Regulatory frameworks and landscape-changing policies for the planning of strategies are intended for climate change adaptation to improve measures to reduce climate change and its impacts (Organization, 2014). Land-use strategies, management, and planning play a vital role in ensuring the mitigations and also providing keys for network infrastructure, along with the protection of ecological services and systems (Lwasa & Njenga, 2012). Finally, it concluded that the practices involved in adaptation should be aligned to reduce the risk socially, economically, and even naturally.

In response to urban resilience, the cities are more exposed to risk, urbanization, economic activity, the concentration of people, and infrastructure development, causing an increase in the risk of exposure. The intensity of natural hazards and the increasing frequency associated with the onset of climate change intensify risk exposure in the Pacific region (Bank, 2013).

The term resilience planning is getting more importance in the urban world, which brings scientific, technical, and knowledge into the decision-making process. This planning of resilience is based on iterative, inclusive, and integrated processes that will reduce the complexity of rapid urban growth and uncertainty caused by climate change.

Mitigation, Sustainable Development Goal 12, and GHG Emission

The impact of greenhouse gas (GHG) on urban areas by emission from the industry is estimated to range from 37 to 49% globally (Seto et al., 2014). The mitigation issues are the major issues for contemporary urban planning, even in city areas of the developed world (Seto et al., 2017). The developed nations make different strategies to decrease the impacts of greenhouse gases; furthermore, low-income cities face different challenges (Parry, 2009). However, in the case of low-income countries, the growth of population and urbanization increased the issues and service provision and spatial planning. To achieve the UN's sustainable goals, it is very necessary to focus on the sectors relevant to mitigation in urban areas, including accessibility and transport, spatial planning, renewable energy, and waste management to evaluate existing policy measures to minimize the burden and provision of benefits.

The low-income cities include some mitigations in city action plans. All of these actions have been adopted and implemented to make services profitable same even in low-income countries as shown in Table 4, but still, these services are not easily available for poor income and push them into risk exposure areas. This impact generates a new term called "climatic gentrification" (Florida, 2018).

The enhancement of public spaces, transportation networks, and road structures is also a part of this mitigation action plan as it brings goods to low-income houses. It is the best action plan to bring the poor population to employment areas that improved their socio-economic profile, which is transit-oriented development (TOD) relating to goal 17. However, it also brings another problem of decreased housing affordability.

Mitigation strategies in transportation mostly affect low-income countries, where the roads lack transportation networks and policies. For example, the registering fees on emission rates also affect the poor-income drivers rather than the owners of the vehicles, as drivers with low income have vehicles that pollute more even per mile (Schweitzer & Valenzuela, 2004).

Another strategy in low- to middle-income states to avoid substantial emissions of greenhouse gases is a waste management business that is waste gathering and recycling and reuse. However, it may face some resistance, from some other authorities, particularly when waste management and collection become an open priority,

Table 4 In what way climate change relates to gender equality in cities

Consequence/distribution-related equity concerns	Adaptation
Equity concerns for gender equality solutions	
Concerns about gender equity are impacted by the primary and secondary effects of climate change on women and girls; e.g., via 1. Heat: women are more affected by heat stress and its effects than males 2. Floods/landslides/storm surges decrease or eliminate the provision of fresh water, sanitary facilities, and drainage in urban areas, particularly among women (a) Loss of revenue from local-based actions and, in many cases, water-based economic activities such as laundry, cleaning, or preparing food goods, particularly in informal settlements (b) Face malnutrition or low-calorie food, particularly elder women and young girls, as a result of gendered diet hierarchies and a reduction in food quantity or a rise in food prices	• Decrease women's risk to vulnerabilities • Expand women's financial prospects and home enterprises • Worth and appropriately compensate care labor and female revenue alternatives • Support methods with suitable financial and institutional maintenance for women in low-income and middle-income countries Mitigation • Understand and avoid gender-based violence, harassment and corruption over gender base mitigation activities, for example, in transport • Prevent a disproportionate impact of transportation policies on girls and women, e.g., by way of costs and access to transportation, as women and girls make more trips and change more frequently than men Adaptation and mitigations • Participation of women in governance, positions in public practices and indigenous climate change policymaking, as females are acknowledged as crucial change agents • Growth of women's participation in local, regional, and international political decision-making, such as the UNFCCC
Process-related equity concerns System equality problems in cities are linked to various gender roles for men and women, and hence varied participation in climate change policy processes	Adaptation and mitigation • Gender roles and women's essentials into consideration in adaptation and mitigation strategies • Challenge conventional gender characters and acknowledge and stable work in men and women • Value and enable paid revenue choices for women when implementing adaptation and mitigation options

Notes: LIC, MIC, and HIC low-income, middle-income, and high-income countries
Adapted from Reckien et al. (2017)

the waste pickers shall be displaced, irrespective of their positive contribution to the environment for business as well as for the social impacts of displacement. Anyhow, in some developed countries like "Uganda" and some cities like "Kampala," contractual arrangements have been set for waste collectors to provide waste management services.

Finally, for climate change mitigation plans some schemes like renewable energy and prevention of greenhouse gases linked with SDG 7, inequality may also increase when the burden of investment is placed on the low-income household as they have larger friction for their energy income and other costs in the household. There should be some well-defined strategies for equality for low- and high-income country residents that are of key importance for sustainable development.

Conclusions and Policy Recommendations

This study examines the association between equality and equity with climate change concerning urban areas to access the level of the United Nations' sustainable goals and attempts to provide evidence of how low-income residents and women get affected by climate change impacts concerning adaptation and mitigations. The study found that gender-sensitive ways of poverty in an urban world responding to climate change, adaptation, and mitigation are the promising corridor that contributes simultaneously to meeting United Nations' sustainability goals.

Climate change has a different level of risk exposure, from preparedness and its impacts. Climate change causes natural disasters, where poor urban are the most vulnerable to exposure, creating a challenge for UN goal 1 creating poverty in the form of property loss, agriculture loss, and environmental disruption. Through adaptation and mitigation policies, women, children, and older and low-income residents get affected most by climate change. Moreover, it is well documented, in the case of low-income and even high-income countries poverty and woman are associated with each other to state climate change and its impacts since the mid-twentieth century. The role of gender and climate change impacts on gender, particularly women, are countless and warranted. However, "Gender and Climate Change" must include "Men's and Climate Change" while discussing vulnerability issues to bridge the gap in previous literature. For instance, men are perhaps more vulnerable to the effects and consequences of climate change due to their work environment. The majority of rescue workers are men, and the dramatic increase in wildfires linked to rising temperatures as well as the predominantly male rescue and reconstruction work connected to disasters like flooding from rising sea levels have a disproportional impact on men.

However, large social security transfers and extensive welfare measures are needed to reduce poverty as suggested by the studies. The findings of this research provide fruitful insights after critical views of past literature and related documents from UN conferences and reports associated with climate change and equity. However, climatic justice and climate change need further research to fill the gap to support the decision-making process. This study also provides some policy guidelines and directions to study the impacts of climate change on gender and equity concerns in developed and developing countries. This equity is related to health, coastal flooding, precipitation, social justice and injustice, adaption, and mitigation as described above to achieve SDG 3 related to good health and well-being, goal 5 related to gender equality, goal 10 reduce inequality, goal 13 climate change, and 17 participation to achieve all UN-SDG without any social, political, or regional discrepancy.

In urban areas, the policies related to adaptation and mitigation are not to increase inequalities. For social justice, there is a need to focus on the low-income population, particularly women, but not limited to poor men, living in poverty, and also reduce the impacts of climate change on the most susceptible groups. Previous research suggests that due to the impacts of climate change poor women are disproportionately affected. However, few adaptations and mitigation policies respect their concern and directly benefit them.

For adaptation measures, it is necessary to understand the infrastructure and service deficiencies in low-income urban populations; it also includes good governance options and builds up institutions, including the finance for necessary actions. It plays a key role to guarantee all gender participation associated with SDG 5, in decision-making policies and processes, to make them active in governing climate change as suggested by the study.

For adoptions and mitigation strategies in low-income neighborhoods and ethnic groups, the spatial planners make a possible attempt to lower the negative impacts of compact city planning and made possible solutions to lower the rents of households and cap accommodation prices by different social policies to achieve SDG 11. Tables 2 and 3 elaborate further recommendations regarding the equity concern for gender equality for climate change and the impacts of climate change adaptations and its mitigations in urban areas.

References

Abbasi, S. S., Anwar, M. Z., Habib, N., Khan, Q., & Waqar, K. (2019). Identifying gender vulnerabilities in context of climate change in Indus basin. *Environmental Development, 31*, 34–42. https://doi.org/10.1016/j.envdev.2018.12.005

Abbass, K., Qasim, M. Z., Song, H., Murshed, M., Mahmood, H., & Younis, I. (2022). A review of the global climate change impacts, adaptation, and sustainable mitigation measures. *Environmental Science and Pollution Research, 29*(28), 42539–42559. https://doi.org/10.1007/s11356-022-19718-6

Addo-Atuah, J., Senhaji-Tomza, B., Ray, D., Basu, P., Loh, F.-H., & Owusu-Daaku, F. (2020). Global health research partnerships in the context of the sustainable development goals (SDGs). *Research in Social and Administrative Pharmacy, 16*(11), 1614–1618. https://doi.org/10.1016/j.sapharm.2020.08.015

Adelekan, I. O. (2019). Urban dynamics, everyday hazards and disaster risks in Ibadan, Nigeria. *Environment and Urbanization, 32*(1), 213–232. https://doi.org/10.1177/0956247819844738

Adnan, M., Xiao, B., Bibi, S., Xiao, P., Zhao, P., & Wang, H. (2024). Addressing current climate issues in Pakistan: An opportunity for a sustainable future. *Environmental Challenges, 15*, 100887. https://doi.org/10.1016/j.envc.2024.100887

Adom, P. K. (2024). The socioeconomic impact of climate change in developing countries over the next decades: A literature survey. *Heliyon, 10*(15), e35134. https://doi.org/10.1016/j.heliyon.2024.e35134

Ajibade, I., McBean, G., & Bezner-Kerr, R. (2013). Urban flooding in Lagos, Nigeria: Patterns of vulnerability and resilience among women. *Global Environmental Change, 23*(6), 1714–1725. https://doi.org/10.1016/j.gloenvcha.2013.08.009

Akanda, A. S., & Hossain, F. (2012). The climate-water-health nexus in emerging megacities. *EOS Transactions, 93*, 353–354. https://doi.org/10.1029/2012EO370001

Al-Humaiqani, M. M., & Al-Ghamdi, S. G. (2022). The built environment resilience qualities to climate change impact: Concepts, frameworks, and directions for future research. *Sustainable Cities and Society, 80*, 103797. https://doi.org/10.1016/j.scs.2022.103797

Amengual, A., Homar, V., Romero, R., Alonso, S., & Ramis, C. (2012). A statistical adjustment of regional climate model outputs to local scales: Application to Platja de Palma, Spain. *Journal of Climate, 25*, 3.

Anser, M. K., Yousaf, S. U., Usman, B., Azam, K., Bandar, N. F. A., Jambari, H., & Zaman, K. (2023). Beyond climate change: Examining the role of environmental justice, agricultural mechanization, and social expenditures in alleviating rural poverty. *Sustainable Futures, 6*, 100130. https://doi.org/10.1016/j.sftr.2023.100130

Appiah, D. O., Asante, F., Antwi-Boadi, L., & Serbeh, R. (2022). Elderly smallholder farmers' perceptions of and adaptation to climate variability and change in rural Ghana. *Working with Older People, 26*(2), 97–108. https://doi.org/10.1108/WWOP-09-2021-0052

Baker, J. L. (2012). *Climate change, disaster risk, and the urban poor*. The World Bank.

Bank, A. D. (2013). Moving from risk to resilience: Sustainable urban development in the pacific. In *Risk reduction and adaptation measures to promote more sustainable urban development.* Retrieved from https://www.adb.org/publications/moving-risk-resilience-sustainable-urban-development-pacific

Baraj, B., Mishra, M., Sudarsan, D., Silva, R. M. D., & Santos, C. A. G. (2024). Climate change and resilience, adaptation, and sustainability of agriculture in India: A bibliometric review. *Heliyon, 10*(8), e29586. https://doi.org/10.1016/j.heliyon.2024.e29586

Barrios, S., Bertinelli, L., & Strobl, E. (2006). Climatic change and rural–urban migration: The case of sub-Saharan Africa. *Journal of Urban Economics, 60*(3), 357–371. https://doi.org/10.1016/j.jue.2006.04.005

Bartlett, A., Smith, G., & King, M. (2009). The response of mental health professionals to clients seeking help to change or redirect same-sex sexual orientation. *BMC Psychiatry, 9*(1), 11. https://doi.org/10.1186/1471-244X-9-11

Bonaiuto, M., Fornara, F., Ariccio, S., Ganucci Cancellieri, U., & Rahimi, L. (2015). Perceived residential environment quality indicators (PREQIs) relevance for UN-HABITAT City prosperity index (CPI). *Habitat International, 45*, 53–63. https://doi.org/10.1016/j.habitatint.2014.06.015

Bulkeley, H., Edwards, G. A. S., & Fuller, S. (2014). Contesting climate justice in the city: Examining politics and practice in urban climate change experiments. *Global Environmental Change, 25*, 31–40. https://doi.org/10.1016/j.gloenvcha.2014.01.009

Bürgi, M., Hersperger, A., & Schneeberger, N. (2004). Driving forces of landscape change - Current and new directions. *Landscape Ecology, 19*, 857–868. https://doi.org/10.1007/s10980-005-0245-3

Colleen Callahan, Daniel Coffee, DeShazo, J. R., & González, S. R. (2021). *Climate equity. Making Justice40 a reality for frontline communities.* Retrieved from https://innovation.luskin.ucla.edu/environmental-equity/climate-equity/

Campbell, S., Remenyi, T. A., White, C. J., & Johnston, F. H. (2018). Heatwave and health impact research: A global review. *Health & Place, 53*, 210–218. https://doi.org/10.1016/j.healthplace.2018.08.017

Carcellar, N., Co, J. C. R., & Hipolito, Z. O. (2011). Addressing disaster risk reduction through community-rooted interventions in the Philippines: Experience of the homeless People's Federation of the Philippines. *Environment and Urbanization, 23*(2), 365–381. https://doi.org/10.1177/0956247811415581

Change, U. N. F. C. O. C. (2022). *Outcomes of the Glasgow climate change conference*. United Nation Climate Change. Retrieved from https://unfccc.int/conference/glasgow-climate-change-conference-october-november-2021

Covich, A. P. (2013). Energy flow and ecosystems. In S. A. Levin (Ed.), *Encyclopedia of biodiversity* (2nd ed., pp. 237–249). Academic Press.

Dankelman, I., & Jansen, W. (2010). Gender, environment and climate change: Understanding the linkages. *Journal of Experimental Psychology-human Perception and Performance*. https://doi.org/10.4324/9781849775274

David, E., & Enarson, E. (2012). The women of Katrina: How gender, race, and class matter in an American disaster.

Development, T. I. I. (2016). *The new urban agenda*. Cities Alliance. Retrieved from https://www.citiesalliance.org/sites/default/files/NUA-iied.pdf

Dodman, D., & Mitlin, D. (2013). Challenges for community-based adaptation: Discovering the potential for transformation. *Journal of International Development, 25*(5), 640–659. https://doi.org/10.1002/jid.1772

Dutta, P., & Dutta, A. (2021). Impact of external assurance on corporate climate change disclosures: New evidence from Finland. *Journal of Applied Accounting Research, 22*(2), 252–285. https://doi.org/10.1108/JAAR-08-2020-0162

Eastin, J. (2018). Climate change and gender equality in developing states. *World Development, 107*, 289–305. https://doi.org/10.1016/j.worlddev.2018.02.021

Eiser, J., Bostrom, A., Burton, I., Johnston, D., McClure, J., Paton, D., & White, M. (2012). Risk interpretation and action: A conceptual framework for responses to natural hazards. *International Journal of Disaster Risk Reduction, 1*, 5–16. https://doi.org/10.1016/j.ijdrr.2012.05.002

Eneji, C.-V. O., Onnoghen, N. U., Acha, J. O., & Diwa, J. B. (2021). Climate change awareness, environmental education and gender role burdens among rural farmers of northern Cross River state, Nigeria. *International Journal of Climate Change Strategies and Management, 13*(4/5), 397–415. https://doi.org/10.1108/IJCCSM-06-2020-0070

Fahy, A. (2021). *What is climate change resilience? Storey about climate*. Retrieved from https://www.concern.net/news/what-climate-change-resilience

Fernandez Milan, B., & Creutzig, F. (2015). Reducing urban heat wave risk in the 21st century. *Current Opinion in Environmental Sustainability, 14*, 221–231. https://doi.org/10.1016/j.cosust.2015.08.002

Florida, R. (2018). *'Climate gentrification' will deepen urban inequality. A new study investigates the intersection of climate change and real estate, and finds that higher elevations bring higher values*. Retrieved from https://www.bloomberg.com/news/articles/2018-07-05/-climate-gentrification-will-deepen-urban--inequality

Fuchs, R. (2010). *Cities at risk: Asia's coastal cities in an age of climate change*. Asia Pacific Issues.

Goodrich, C. G., Prakash, A., & Udas, P. B. (2019). Gendered vulnerability and adaptation in Hindu-Kush Himalayas: Research insights. *Environmental Development, 31*, 1–8. https://doi.org/10.1016/j.envdev.2019.01.001

Güneralp, B., Güneralp, I., & Liu, Y. (2015). Changing global patterns of urban exposure to flood and drought hazards. *Global Environmental Change, 31*, 217–225. https://doi.org/10.1016/j.gloenvcha.2015.01.002

Gupta, A. (2016). Chapter 1 - Climate change and Kyoto protocol: An overview. In V. Ramiah & G. N. Gregoriou (Eds.), *Handbook of environmental and sustainable finance* (pp. 3–23). Academic Press.

Hák, T., Janoušková, S., & Moldan, B. (2016). Sustainable development goals: A need for relevant indicators. *Ecological Indicators, 60*, 565–573. https://doi.org/10.1016/j.ecolind.2015.08.003

Huynh, P. T. A., Le, N. D., Le, S. T. H., & Tran, T. N. (2021). Adaptive livelihood strategies among small-scale fishing households to climate change-related stressors in central coast Vietnam. *International Journal of Climate Change Strategies and Management, 13*(4/5), 492–510. https://doi.org/10.1108/IJCCSM-04-2020-0034

Jabeen, T., & Jabeen, S. (2018). Ideals of human rights and socioeconomic realities: The larger context of Pakistan's child-protection policy. *Journal of Human Rights, 17*(1), 44–57. https://doi.org/10.1080/14754835.2016.1233807

Kallbekken, S., Sælen, H., & Underdal, A. (2014). *Equity and spectrum of mitigation commitments in the 2015 agreement*. Nordisk Ministerråd.

Kawamoto, M., & Kanie, N. (2020). Engaging business: The UN sustainable development goals and climate change. In S. Hori, Y. Takamura, T. Fujita, & N. Kanie (Eds.), *International development and the environment: Social consensus and cooperative measures for sustainability* (pp. 47–54). Springer.

Kern, K., & Bulkeley, H. (2009). Cities, Europeanization and multi-level governance: Governing climate change through transnational municipal networks. *Journal of Common Market Studies, 47*, 309–332. https://doi.org/10.1111/j.1468-5965.2009.00806.x

Khan, N. U., Zhongyi, P., Ullah, A., & Mumtaz, M. (2024). A comprehensive evaluation of sustainable mineral resources governance in Pakistan: An analysis of challenges and reforms. *Resources Policy, 88*, 104383. https://doi.org/10.1016/j.resourpol.2023.104383

Klinsky, S., & Dowlatabadi, H. (2009). Conceptualizations of justice in climate policy. *Climate Policy, 9*(1), 88–108. https://doi.org/10.3763/cpol.2008.0583b

Komori, J., Koike, T., Yamanokuchi, T., & Tshering, P. (2012). Glacial lake outburst events in the Bhutan Himalayas. *Global Environmental Research, 16*, 59–70.

Kousky, C., Kunreuther, H., LaCour-Little, M., & Wachter, S. (2020). Flood risk and the U.S. housing market. *Journal of Housing Research, 29*(sup1), S3–S24. https://doi.org/10.1080/10527001.2020.1836915

Kovats, S., & Akhtar, R. (2008). Climate, climate change and human health in Asian cities. *Environment and Urbanization, 20*(1), 165–175. https://doi.org/10.1177/0956247808089154

Krishna, A., Sriram, M. S., & Prakash, P. (2014). Slum types and adaptation strategies: Identifying policy-relevant differences in Bangalore. *Environment and Urbanization, 26*(2), 568–585. https://doi.org/10.1177/0956247814537958

Kuh, K. F. (2018). The law of climate change mitigation: An overview. In D. A. Dellasala & M. I. Goldstein (Eds.), *Encyclopedia of the anthropocene* (pp. 505–510). Elsevier.

Laaidi, K., Zeghnoun, A., Dousset, B., Bretin, P., Vandentorren, S., Giraudet, E., & Beaudeau, P. (2012). The impact of heat islands on mortality in Paris during the august 2003 heat wave. *Environmental Health Perspectives, 120*(2), 254–259. https://doi.org/10.1289/ehp.1103532

Lawhon, M. (2014). Diana Mitlin and David Satterthwaite 2013: Urban poverty in the global south: Scale and nature. *International Journal of Urban and Regional Research, 38*(1), 370–371. https://doi.org/10.1111/1468-2427.12137_6

Leal Filho, W., Lovren, V. O., Will, M., Salvia, A. L., & Frankenberger, F. (2021). Poverty: A central barrier to the implementation of the UN sustainable development goals. *Environmental Science & Policy, 125*, 96–104. https://doi.org/10.1016/j.envsci.2021.08.020

Li, Y., Qin, Y., Ma, L., & Pan, Z. (2020). Climate change: Vegetation and phenological phase dynamics. *International Journal of Climate Change Strategies and Management, 12*(4), 495–509. https://doi.org/10.1108/IJCCSM-06-2019-0037

Lwasa, S., & Njenga, C. (2012). *Reappraising urban planning and urban sustainability in East Africa*.

McGranahan, G., Balk, D., & Anderson, B. (2007). The rising tide: Assessing the risks of climate change and human settlements in low elevation coastal zones. *Environment and Urbanization, 19*(1), 17–37. https://doi.org/10.1177/0956247807076960

Metz, B. (2000). International equity in climate change policy. *Integrated Assessment, 1*(2), 111–126. https://doi.org/10.1023/A:1019171513507

Mohammad, A. J. (2022). *Decade of action. Decade of action to deliver the global goals*. Retrieved from https://www.un.org/sustainabledevelopment/decade-of-action/

Morton, S., Pencheon, D., & Bickler, G. (2019). The sustainable development goals provide an important framework for addressing dangerous climate change and achieving wider public health benefits. *Public Health, 174*, 65–68. https://doi.org/10.1016/j.puhe.2019.05.018

Nabangchang, O., Allaire, M., Leangcharoen, P., Jarungrattanapong, R., & Whittington, D. (2015). Economic costs incurred by households in the 2011 greater Bangkok flood. *Water Resources Research, 51*(1), 58–77. https://doi.org/10.1002/2014WR015982

Nation, U. (2015). *The 17 goals*. Retrieved from https://sdgs.un.org/goals

Nation, U. (2017). *UN climate change conference - November 2017. UN climate Change conference*. Retrieved from https://unfccc.int/process-and-meetings/conferences/past-conferences/un-climate-change-conference-november-2017/about/un-climate-change-conference-november-2017

Nation, U. (2022). *World urban forum*. Retrieved from https://wuf.unhabitat.org/

Nations, U. (2016). *The Paris agreement. Legally binding international treaty on climate change*. Retrieved from https://unfccc.int/process-and-meetings/the-paris-agreement/the-paris-agreement

Nations, U. (2018). *Millennium development goals (MDGs). Millennium development goals (MDGs)*. World Health Organization.

Nitschke, N., Allan, E., Zwölfer, H., Wagner, L., Creutzburg, S., Baur, H., & Weisser, W. W. (2017). Plant diversity has contrasting effects on herbivore and parasitoid abundance in Centaurea jacea flower heads. *Ecology and Evolution, 7*(22), 9319–9332. https://doi.org/10.1002/ece3.3142

Nuru, J. T., Rhoades, J. L., & Gruber, J. S. (2021a). Evidence of adaptation, mitigation, and development co-benefits of solar mini-grids in rural Ghana. *Energy and Climate Change, 2*, 100024. https://doi.org/10.1016/j.egycc.2021.100024

Nuru, J. T., Rhoades, J. L., & Gruber, J. S. (2021b). The socio-technical barriers and strategies for overcoming the barriers to deploying solar mini-grids in rural islands: Evidence from Ghana. *Technology in Society, 65*, 101586. https://doi.org/10.1016/j.techsoc.2021.101586

Organization, W. S. E. D. (2014). *Ensuring women's access and influence on climate change policy*. Retrieved from https://wedo.org/ensuring-womens-access-and-influence-on-climate-change-policy-2/

Parry, M. (2009). Climate change is a development issue, and only sustainable development can confront the challenge. *Climate and Development, 1*, 5–9. https://doi.org/10.3763/cdev.2009.0012

Pérez-Molina, E., Sliuzas, R., Flacke, J., & Jetten, V. G. (2017). Developing a cellular automata model of urban growth to inform spatial policy for flood mitigation: A case study in Kampala, Uganda. *Computers, Environment and Urban Systems, 65*, 53–65. https://doi.org/10.1016/j.compenvurbsys.2017.04.013

Qiang, N. (2014). *Programme world council 2014. United cities and local government*. Retrieved from https://www.uclg.org/en/node/22781.

Rahman, A.-U., & Khan, A. (2013). Analysis of flood causes and associated socio-economic damages in the Hindu Kush region. *Natural Hazards, 59*, 1239–1260.

Rankoana, S. A. (2020). Climate change impacts on water resources in a rural community in Limpopo province, South Africa: a community-based adaptation to water insecurity. *International Journal of Climate Change Strategies and Management, 12*(5), 587–598. https://doi.org/10.1108/IJCCSM-04-2020-0033

Rao, M. A., Berry, R., Gonsalves, A., Hastak, Y., Shah, M., & Roeser, R. W. (2013). Globalization and the identity remix among urban adolescents in India. *Journal of Research on Adolescence, 23*(1), 9–24. https://doi.org/10.1111/jora.12002

Reckien, D., Creutzig, F., Fernandez, B., Lwasa, S., Tovar-Restrepo, M., McEvoy, D., & Satterthwaite, D. (2017). Climate change, equity and the sustainable development goals: An urban perspective. *Environment and Urbanization, 29*(1), 159–182. https://doi.org/10.1177/0956247816677778

Reckien, D., Flacke, J., Dawson, R., Heidrich, O., Olazabal, M., Foley, A., & Pietrapertosa, F. (2014). Climate change response in Europe: What's the reality? Analysis of adaptation and mitigation plans from 200 urban areas in 11 countries. *Climatic Change, 122*, 331–340. https://doi.org/10.1007/s10584-013-0989-8

Reduction, U. N. O. F. D. R. (2009). *2009 UNISDR terminology on disaster risk reduction*. United Nations Office for Disaster Risk Reduction. Retrieved from https://www.undrr.org/publication/2009-unisdr-terminology-disaster-risk-reduction#:~:text=30%20p.,authorities%2C%20practitioners%20and%20the%20public

Revi, A., Satterthwaite, D., Aragón-Durand, F., Corfee-Morlot, J., Kiunsi, R., Pelling, M., & Solecki, W. (2014). Urban areas in climate change 2014: Impacts, adaptation, and vulnerability. In *Part A: Global and sectoral aspects* (pp. 535–612). IPCC.

Revi, A., Satterthwaite, D., Aragón-Durand, F., Corfee-Morlot, J., Kiunsi, R. B. R., Pelling, M., & Sverdlik, A. (2014). Towards transformative adaptation in cities: The IPCC'S fifth assessment. *Environment and Urbanization, 26*(1), 11–28. https://doi.org/10.1177/0956247814523539

Rocha, J., Oliveira, S., Viana, C. M., & Ribeiro, A. I. (2022). Chapter 8 - Climate change and its impacts on health, environment and economy. In J. C. Prata, A. I. Ribeiro, & T. Rocha-Santos (Eds.), *One health* (pp. 253–279). Academic Press.

Satterthwaite, D., Revi, A., Aragón-Durand, F., Corfee-Morlot, J., Kiunsi, R., Pelling, M., & Solecki, W. (2014). Chapter 8 in climate change 2014: Impacts, adaptation and vulnerability. In *One health* (p. 113). Academic Press.

Schlosberg, D., & Collins, L. B. (2014). From environmental to climate justice: Climate change and the discourse of environmental justice. *WIREs Climate Change, 5*(3), 359–374. https://doi.org/10.1002/wcc.275

Schuster, M. A., McGlynn E, Brook, R. H., & Brook, R. H. (n.d.). How good is the quality of health care in the United States?

Schweitzer, L., & Valenzuela, A. (2004). Environmental injustice and transportation: The claims and the evidence. *Journal of Planning Literature, 18*, 383–398. https://doi.org/10.1177/0885412204262958

Seto, K. C., Dhakal, S., Bigio, A., Blanco, H., Delgado, G., Dewar, D., & Ramaswami, A. (2014). *Human settlements, infrastructure and spatial planning*.

Seto, K. C., Golden, J. S., Alberti, M., & Turner, B. L. (2017). Sustainability in an urbanizing planet. *Proceedings of the National Academy of Sciences, 114*(34), 8935. https://doi.org/10.1073/pnas.1606037114

Singh, J., Grann, M., & Fazel, S. (2010). A comparative study of violence risk assessment tools: A systematic review and metaregression analysis of 68 studies involving 25,980 participants. *Clinical Psychology Review, 31*, 499–513. https://doi.org/10.1016/j.cpr.2010.11.009

Sun, S., Dai, T.-L., Wang, Z.-Y., Chou, J.-M., Chao, Q.-C., & Shi, P.-J. (2021). Projected increases in population exposure of daily climate extremes in eastern China by 2050. *Advances in Climate Change Research*. https://doi.org/10.1016/j.accre.2021.09.014

Tompkins, C., Klepousniotou, E., & Scott, A. (2013). *Nature and assessment of right hemisphere disorders*.

Tovar-Restrepo, M., & Irazábal, C. (2013). Indigenous women and violence in Colombia: Agency, autonomy, and territoriality. *Latin American Perspectives, 41*(1), 39–58. https://doi.org/10.1177/0094582X13492134

Tuckett, R. P. (2009). Chapter 1 - The role of atmospheric gases in global warming. In T. M. Letcher (Ed.), *Climate change* (pp. 3–19). Elsevier.

Ullah, S., & Lin, B. (2024). Natural resources, renewable energy-environment nexus for Pakistan: A policy perspective. *Resources Policy, 90*, 104788. https://doi.org/10.1016/j.resourpol.2024.104788

UN-Habitat. (2010). *UN-Habitat for better future*. United Nations Human Settlements Programme. Retrieved from https://unhabitat.org/state-of-the-worlds-cities-20102011-cities-for-all-bridging-the-urban-divide

Wan, G., Zhang, X., & Zhao, M. (2022). Urbanization can help reduce income inequality. *NPJ Urban Sustainability, 2*(1), 1. https://doi.org/10.1038/s42949-021-00040-y

Wernberg, T., Bennett, S., Babcock, R. C., de Bettignies, T., Cure, K., Depczynski, M., & Wilson, S. (2016). Climate-driven regime shift of a temperate marine ecosystem. *Science, 353*(6295), 169–172. https://doi.org/10.1126/science.aad8745

White-Newsome, J., Sánchez, B., Parker, E., Dvonch, J., Zhang, Z., & O'Neill, M. (2011). Assessing heat-adaptive behaviors among older, urban-dwelling adults. *Maturitas, 70*, 85–91. https://doi.org/10.1016/j.maturitas.2011.06.015

Wu, Y., Xi, X., Tang, X., Luo, D., Gu, B., Lam, S. K., & Chen, D. (2018). Policy distortions, farm size, and the overuse of agricultural chemicals in China. *Proceedings of the National Academy of Sciences, 115*(27), 7010–7015. https://doi.org/10.1073/pnas.1806645115

Zhang, K., Chen, T. H., & Begley, C. E. (2020). *Impact of the 2011 heat wave on mortality and emergency department visits in Houston, Texas*. (1476-069X (electronic)).

The Neighborhoods Services and Their Impacts on Property Values in Urban Area: A Review by Hedonic Pricing Model

Asad Aziz and Muhammad Mushahid Anwar

Abstract

The proximity to neighborhoods services is a valuable asset for land and property in an urban area. Neighborhoods services may be the natural corridor of trees as the trail of environmental amenity in the form of green trees canopy, urban green or forest in a region and parks, open street network, proximity to the major roads, market or a worship place, size of a building, age of a house, numbers of rooms in a building, and location of a property. The assessment of the impact of neighborhoods services is still unknown in many of the developing countries. However, most of the studies use the Hedonic Pricing Model for their evaluation by using regression analyses. This is proposed to study and review the past literature to examine the impacts of neighborhoods services on property values. The research concluded that properties near a park, worship place, market area, and a house with wide street networks have more value compared to those located at a distance. This phenomenon is mainly due to the proximity to the neighborhoods services. The research provides valuable insights into property buyers, investors, and government officials for urban planning and land-use development for urban area.

Keywords

Neighborhoods services · Hedonic pricing model · Property · Proximity · Planning · Location · Linear multi-regressions analysis · Price · Urban development · Values

Introduction

Access or proximity to a neighborhood's service, for example, access to a major road, market, worship place, and proximity to the road, has influential impacts on property values in an urban area. These services are recognized as environmental justice for urban dwellers in a planned urban area. Much less attention has been paid to these services close to the public places and residences in terms of proximity to these services due to rapid and unplanned urbanization and city expansion. Cities are becoming more dependent on natural environment and ecosystem beyond their limits. However, the proposed study will provide some findings by examining the impacts of neighborhoods services on property sale value. This evaluation of these services is mostly done by Hedonic Pricing Model, which use multiple regression analyses for their economic assessment.

A. Aziz (✉) · M. M. Anwar
Department of Geography, University of Gujrat, Gujrat, Punjab, Pakistan
e-mail: asadaziz@uog.edu.pk; mushahid.anwar@uog.edu.pk

M. M. Anwar, A. Aziz (eds.), *Sustainable Landscape for Urban Resilience*, Sustainable Development Goals Series, https://doi.org/10.1007/978-3-031-99836-2_5

In developing countries, the previous data of past studies describes a remarkable increase in human population and migration around the world, which is growing rapidly. The urban sprawl in recent decades disturbs the urban life quality and creates crucial issues such as congested urban infrastructure, road network, overcrowded public spaces, environmental degradation, air, land, and noise pollution. The quality of life has different elements in which green spaces are important; therefore, this case study achieves the objects of research and provides an overview and tools for better land-use planning and sales comparison methods.

In pursuit of a sustainable management plan and establishment of urban environmental sustainability, the concerning authorities such as the local government should maintain a database related to ecological, economic, and social values. This database is very helpful for management, mitigation plan for strategic circumstances, and natural conservations. The key role for conservation is deeply related to "if you built, you will have to maintain it." In spite of these challenges, a number of problems and issues are faced by green spaces currently such as poor planning, lack of investment, natural environmental problems, social problem, and instability of political infrastructure. Other intermediate issues and factors including a poor use of technology and scientific means also create challenges for these neighborhoods services. The research proves that such studies are the forceful tools for land-use planning and urban development to model the land use of an urban area.

Review of Literature

The word "Hedonic" has a Greek origin, meaning the "pleasure" (Prunty, 2016); therefore, Hedonic Pricing means "pay for pleasure." The Hedonic Model is one of the best models used to measure the relative location of property in economic terms (Brereton et al., 2008). This model leads the valuation of the environmental and neighborhoods services in an area (Mayora et al., 2009). Green spaces are a component of green infrastructure not only supporting the lifestyle but also ensuring the quality of life in urban areas (WHO, 2017). The green spaces have a greater impact on microclimatic condition in urban areas (Georgi & Dimitriou, 2010). They reduce the heat effects in an urban area and also provide a comfortable outdoor environment (Lafortezzaab et al., 2009). A property has high values when located closer to the parks and roads (Sandera & Polasky, 2009). The green spaces always have a valuable consideration economic development. In the last decades, the demand for green environment has been increasing rapidly due to urbanization.

The Hedonic Model is one of the best models to measure the relative location in economic terms. This is the leading method of the evaluation of environment. Neighborhoods services are a component of green infrastructure not only promoting the life style, also ensuring the quality of life naturally (WHO, 2017). The neighborhoods services have a greater impact on microclimatic condition in urban areas. The neighborhoods services not only reduced the heat effects in an urban area but also provide the comfortable outdoor environment. The prices of the properties increase when located closer to the parks, neighborhoods services, or natural amenity. Traditionally the housing prices are estimated by the supply and demand comparison; the housing is the fundamental need of human beings. An increasing rate of unemployment will increase the homelessness as well as increase the rent of the housing (Prunty, 2016). Traditional methods have drawbacks because of substantial changes happening in urban areas.

The hedonic index uses the individual property and individual location of a house at a time. In property estimations and research of housing marketing, the usage of multiple regression techniques on a large data set is a formal way for analysis of data. The green spaces, natural or man-made, always have a valuable consideration economically and in developmental context. The demand of green environment has been increasing rapidly due to urbanization. Proximity to neighborhoods services has a statistically significant effect on the property values. Neighborhoods services provide public benefits and explore the multiple ways and impacts on economic values even the marginal cost to proximity examined by local to global level from geographically weighted and ordinary least square.

In 1998, China had residential housing reforms, on the basis of the Hedonic Pricing Model that valued the services by green spaces (Diao, 2007). In Europe and North America in the nineteenth century, evidence proved that the relationship between health and green space plays a vital role in social and urban infrastructure improvement (WHO, 2016). Green spaces are the environmental amenity in an urban area (Ham, 2011). Very few studies also show negative impacts of green spaces such as the green buffering of infrastructure or factory called unattractive land use for residency (Panduro & Veie, 2013). For this reason, a multipurpose criterion related to socio-economic, environment, and ecology has been prepared, with category of green spaces and their benefits.

Neighborhoods services provide many services and benefits depending upon the way we use as the neighborhoods services are the cause of leisure opportunities and aesthetic enjoyment. In China, in 1998, the residential housing reforms took place for the services of green spaces that valued the services by neighborhoods services on the basis of the Hedonic Pricing Model. A number of evidences proved that the relationship between health and green space is one of the leading forces for urban parks improvement like in Europe and North America in the nineteenth century (WHO, 2016). In the urban area, neighborhoods services of the environmental amenity (Ham, 2011) showed in economic analysis that it has a deep and prolong effect over property value. But few studies also show some negative effects of green spaces such as the buffering of infrastructure or factory called unattractive land use for residency. A multipurpose criterion in Table 1, along with the category of the neighborhoods services and benefits from them. This study will examine the green spaces not only economically, ecologically, but also environmental and social prospects. Table 1 categorizes the services by neighborhoods services and analyzes the benefits provided by them.

Table 1 Attributes of property values

Green ways category/ services	Values/benefits
Recreations	Facilitate us in nature observations, exercise, jogging, cycling or walking, etc.
Property	Land value increases to get the services of open green spaces
Health/fitness	Physical activities create physical health and terms of secondary health benefits
Transportations	Access to neighborhoods services transportations time and distance
Ecological/ environmental services	Existence of natural species, natural conservations biodiversity, pollution control, and runoff
Visual amenity/ psychological services	Neighbors and passengers get the psychological benefits of open green spaces naturally

The sale property values and neighborhoods services are interlinked with each other not only by economical prospectus, but also in ecological, physiological, as well as recreational. Urban dwellers and visitors use the neighborhoods services their own way like exercise, skating or cycling, playing, or some other physical activities. The planners and policymakers should pay attention to study the environmental amenity in the given region and their impacts on economic vitality. The variables to judge the effect of the Hedonic Model can be grouped into four major classes, which are further divided into more attributes. The major categories are location, structure, neighborhood, and community bases. Table 2 explains the positive or negative effects on the sale value of properties.

The attributes are nothing by themselves, but the usage of attributes related to them enhanced their utilities, so the price also, with the same trend. A house with more attributes will have more demand of a high price. This is the basic function of the Hedonic Pricing Model. The Hedonic Pricing mode analysis is a reflected part of the Multi-Regression Model, which is resulted by the sum of variables and the equation is driven

Table 2 Attributes, with their effects on the values of properties

1. Structural	
Area/size of house	+ve
Number of bedrooms	+ve
Fireplaces, garage, basement	+ve
Age of building	−ve
2. Neighborhood	
Easy access to school and mosque	+ve
Easy access to mosque	+ve
Traffic noise from the road	−ve
Neighborhood quality (air quality, garden)	+ve
3. Community	
Same religion	+ve
Same cast	+ve
Financial status	+ve
Job status	+ve
4. Location	
Distance from neighborhoods services (natural amenity)	−ve
More distance from market/shopping center	−ve
Adjacent to the roadside	+ve

+ve = positive impact on housing price. −ve = negative impact on housing price

by the interactions among multiple variables as stated in the table that is equal to the sum of sale values. Proximity to services is the neighborhood level, and it is associated with sales values property attributes such as the size of the building, lot size, and proximity to open space; understanding the hedonic structure is best for urban planning policies. The activities are very effective to decrease the medical expenditures. The neighborhoods services have more importance in urban areas compared to rural areas because the rural areas are naturally green. Most of the previous data shows different aspects of the impact of neighborhoods services whether it is positive or negative in case of economic significance and privacy concern, respectively. The Hedonic Model tells us the extent of neighborhoods services on property value by comparing the variables based upon the analysis not only with the context of economic prospects also in psychological dimensions as the neighborhoods services provide us a natural amenity by open green vegetation. Neighborhoods services provide a range of positive benefits.

The neighborhoods services have a direct impact on property and economic pricing and increased the extent of economic development. In order to estimate the total effect of neighborhoods services on sale value of property in terms of economic development, it is to some extent a complicated matter; however, most of the studies showed a complicated correlation among different variables like lot size, number of bedrooms, and structure of houses, neighborhood characteristics, and locational attributes. Economic values and benefits of green spaces are estimated not only by the sales values, but also in terms of traveling costs, time to reach at green area, mode of transportations, and distance. Good urbanism not only provides a quality of life, but also enhances the property value along with other features such as characteristics of neighborhoods as such the green ways (Diao, 2007). The different characteristics such as the location of house, size, utilities, indoor/outdoor, and ownership also participate in determining the sale value. The housing index such as the location, facilities, and history determines the degree to which houses vary from each other. The economic valuation of neighborhoods services is supportive for city development and urban planning. By using the knowledge of the importance of neighborhoods services, we can manage the property value and also make contributions to decision-making and city planning.

Types of Neighborhoods Services and Their Functions

Table 3 provides the definition of different types of neighborhoods services and their functions.

Neighborhoods Services and Their Values

(a) Rent values: A property near to green space will have high rent because of proximity to green paces
(b) Traveling cost: Near or adjacent to a green space is, no traveling cost for a visitor or

Table 3 Classification of urban neighborhood services and their functions

1. Park and playground	An area having 2 ha size and no one lives away from the park at a distance of 300 m There should 20 ha parks within 2 km radius in a city	A park can be defined as a green area that has a high level of maintenance, well-defined external, internal, and social accessibility and has a number of recreational and amenity services for visitors and local dwellers. Small lakes, ponds, flowers, garden, and sports activities are the main features of a park or green area
2. Natural green	The natural green are on large landscape near settlement	Sometimes, especially near the edge of city, there is an area with open field, green grass, and open land without human infrastructure treated as nature, which allows people visit the natural landscape
3. Agricultural areas	The seasonal crops or green feeder	The open man land/agricultural area located near houses and settlement, where the local public grow seasonal crops and green feeder
4. Sports fields	A platform for physical activity	The open large open fields or playground that provide facility of playing and the landscape for physical and sports activities. Playgrounds and sports fields are often outlined by trees by linear rays and covered with green grass. These places are sometimes also used for social events
5. Agricultural area	An area of homogeneous landscape	Mostly in rural areas, these places do not have any external, internal, or social accessibility. Just farmers used these places for agricultural activities. Mostly these places have a homogeneous landscape
6. Green buffers	A ray of trees surrounding the factory	The areas that are located around the major roads, industry, or other noisy places often covered by trees directly attached with infrastructure are the green buffer. The main purpose of this buffer is to reduce the impact of noise and air pollution in an urban area. Similarly, the industries often have a green buffer in surroundings, and this area is mainly used by workers and neighborhoods residential for recreational purposes
7. Urban parks	A formal place for recreation very close to residential community	An area in dense urban network, in which locals can spend their free time for leisure and community engagement
8. Accessibility to road	Effective for proximity	Accessibility to major road provides proximity to many services, like transportation, low rent, and easy walk, which will increase the property values

dweller willing to pay to reach the green space.

(c) Business location: A green area that attracts tourists, which increase business of the local community in terms of large sales, tax generation, and business share. A market with green area will have more extent of visitors than a market with no green area (Aram et al., 2019).

(d) Energy cost reduction: The trees and green areas decrease air-conditioning cost by maintaining the air temperature in an urban area.

Environmental Benefits

(a) Biodiversity: Green spaces increase the level of biodiversity in local area by providing ecosystem and environmental services in terms of parks, recreation site, and tourist spots. The public green areas are the assemblage of different plants and animals that are a growing concern of habitat loss and biodiversity (Farinha-Marques et al., 2016).

(b) Erosion control: The trees, shrubs, and grasses prevent erosions and maintain the soil structure and fertility. The studies show the dense green spaces are three times effective than unhealthy green spaces to control nitrogen runoff. Similarly, a lawn with grass is more effective than a wheat field to absorbed rainwater (Heinze, 2011).

(c) Air purification: The green spaces are natural tools for purification of urban air quality on a daily basis, which will control the breathing disorder and high incidence of asthma in an urban area. Along with purification of air, urban trees play a significant role in moderating the temperature in the summer season.

(d) Oxygen generation and carbon reduction: The green area by size 50/50 sq. ft. releases oxygen for a family of four on a daily basis (Heinze, 2011). The trees absorb a large amount of carbon dioxide generated from traffic and industries. The vegetation in the Northern Hemisphere plays a vital role in maintaining the carbon cycle on land and atmosphere and is very sensitive to climate change (Chu et al., 2019). This effect is more dominated in urban ecosystem.

Health/Social Benefits

(a) Recreation: Green spaces are the best platform for recreation providing aesthetics values, platforms for physical and sports activities, etc. The parks create a potential in public design of a city (Luximon et al., 2015). The landscape planning for city population underlines the development and maintenance of green spaces for their number of social and environmental benefits (Kabisch et al., 2015).

(b) Platform for physical activities: The adjacent open spaces in an urban area are stressed due to urbanization (Hanauer & Reid, 2017). Today the structure of houses, especially in urban areas, is very narrow due to overpopulation and dense housing structure. The green spaces are the component of urban landscape that encourage people for outdoor activity and create an ecological function (Zhang et al., 2013). The green spaces in a dense urban area increase the outdoor activities in everyday life (Peschardt et al., 2012). But nowadays, in dense cities the urban parks face problems and limited resources for maintenance (Peschardt & Stigsdotter, 2013).

(c) Health care and stress reduction: For the local urbanities and hospital patients, walking a few minutes on green spaces releases stress. Studies in England show the disparities due to economic level are a much narrow area with ample green spaces (Heinze, 2011). Physiologically, walking on green spaces causes a reduction in problems like fear and anger and creates positive emotions. Similarly, the urban parks and green grassy ground also stable the blood pressure, heartbeat, and body temperature. The studies prove that the social behavior of people living near green spaces give the sense of belonging to each other (Aram et al., 2019).

However, research studies show that green spaces have a wide range of economic, social, educational, and ecological benefits associated with green renewal. Finally, it is concluded that there is a strong need of raising awareness about green spaces and their potential on human life, political attention, reshaping the cultural landscape values and investment to provide resources to green spaces (Dunnett et al., 2002).

Housing quality is a temporal phenomenon because of change in quality of house, age, and material capacity. This model is criticized by many authors due to time instability. Time is a dynamic phenomenon, for instance, Millen (2003) uses alternative repeat-sales model for the identification of change in housing price in early 1983 in Chicago. He found out that increasing distance from the Central Business District (CBD) does not affect the price of houses while the situation was reversed in 1990 when the "CBD" was re-established and the distance to "CBD" affected the housing prices in sales by 8% in total values of the property and houses. His study also explains the variation in time with respect to the demand for houses and property, which increased due to the establishment of "CBD" at that time. Similarly, Corrado et al. (2005) explain that the quality and inflation change over time; therefore, the Hedonic Model fails to capture these values. For example, in the demographic case, an opening of a new job center will attract people to live near, so the location preference in terms of values will change over time. Generally, a number of scholars attempted to explain and explore how the quality of house or property will change due to the addition and subtraction in temporal changes. On the other hand, these attributes will be positive or negative, for example, opening of the new railway station near the housing colony will change values. For a few people, it will be positive like businessmen, while it will be negative for some people like patients due to the noise pollution. For a businessman, the accessibility is important while for a patient noise pollution is a problem. Therefore, the quality change with time will affect the overall value of house or property.

Neighborhoods Services and Their Usage

Neighborhoods services by multipurpose usage enhanced not only the urban infrastructure but also provide a natural sense of diversity, natural conservations, recreation opportunities, economic development, and decreased transportation costs. Location-specific services such as green spaces have effects on life satisfaction directly. The housing price is difficult to measure due to changes in characteristics and quality of the houses. The Hedonic Model has two main components: first, the land owners get benefits by direct observations of open neighborhoods services; second, they derive the benefits by close and easy access purposes. The Hedonic Model and land values are common factors that are not used for the estimations of urban problems. This review study is probably a unique contribution in this sense.

The regression analysis is a technique used to estimate the variables that are related to each other with reasons and result while the regression, where one variable is dependent and more than two or more are independent variable, is called linear regression. The linear multi-regression model is run to judge all of these variables statistically. This model is a predictive model used to explain the relationship among a dependent variable and number of continuous independent variables. The linear multi-regression model is one of the most powerful models to show how a change in single variable affects the property value. The existence of green ways also determines the property value as well as natural corridor that is more valuable than man-made.

Methods

The methodology of the review paper is based on the critical review of past literature related to the property and real estate market. Initially, a total of 1200 past reports, research papers, and real estate values registers were selected. The selection process follows the keywords matching the major objectives of the review. These keywords are property values, property prices, housing market, houses and market, price and values, Hedonic Model, neighborhoods services, and housing structure. These raw data were further filtered by the most related paper with the Hedonic assessment, which provides only 157 documents. After screening further, a more comprehensive review was done to select the closer studies done on Hedonic assessment by using the variables, including structural, locational, environmental, and community attributes. After applying the filter, the remaining documents were only 34. These remaining documents provide a comprehensive data regarding the Hedonic estimation of housing and property prices in the real estate market.

Discussion

The neighborhoods services in urban areas are the natural environmental amenity for public usage, which has a positive impact on multiple approaches, including economic, environmental, and planning and ecological aspects (Wu, 2006). The assessment of neighborhoods services in urban area provides a forceful tool for urban planning and development strategies to ensure the quality of life (Kausar et al., 2024). Every category of neighborhoods services has its own services and benefits, for example, structural, locational, neighborhoods, or community. Most of the studies emphasized the importance of distance of open neighborhoods services on sales values with greater distance has a significant negative impact on property prices and a direct correlation (Ferreira et al., 2018; Metz, 2015), which means more the distance less the values or price. The services provided by the neighborhood's amenities, for example, urban parks or trees canopy near a residency, not only reduced the stress, but also caused reduction in energy competition for cooling and reduced the soil erosion. However, the studies of this issue open the doors for further research.

The open land available for sale and property dealers who deal with sale purchase process should be bound by local government to sell the land portion according to policies provided by the government for land-use planning. For example, the total available land should be divided into groups, and each group should be specified for residential and commercial area, respectively.

The study shows that the impacts of neighborhood services are not only important for health and environment, but also have importance in economic measures like increase in property values, increase in rent, decease in transportations cost when a property is located near the road, and other locational expenditures. It has been analyzed that neighborhoods have a positive impact on society and sales and rental property values (Aziz et al., 2021). A house/property near the road will have more values as well as more rent compared to others located away from the road. However, one of the most significant aspects found in the biological aspect of green spaces is whether they are present in core urban or peri-urban area, and they stabilize and attract the tourists around the city (Cui et al., 2018) (Sangwan et al., 2022). But due to the increasing rate of land requirements for settlement in urban areas, it has caused a decrease in natural land by conversion of land into built-up area. Today, in developing countries land allocation is a challenge because land-use pattern is irregular and proved by a number of studies (Nha, 2017; Schreinemachers & Berger, 2006). For example, one portion of the city has a dense array of hospitals while another portion has not a single one. The environmental and social attributes enhance the locational magnificence that determines the land/property values (Wang et al., 2023). Planners and scholars debate on estimating the impacts of green spaces on property sale values for urban planning.

Conclusion

The search concluded that green spaces, including parks, agriculture fields, open land or land adjacent to roads, wide area, proximity to roads and streets networks, are a fundamental need of man nowadays that ensures quality of life, especially in an urban area. Similarly, the neighborhood services, including the presence of park, accessibility to hospital, market, and location of property near road, increase the land and property values. A property with multiple attributes, including large area, greater number of rooms, and larger structure of covered area, increases the value of the property and attracts property buyers. Every sphere of human life is changing, including social, economic, and ecological dimensions. Meanwhile, the economic evaluation and assessment of these dimensions are necessary for urban planning today, where a number of challenges for urban community have been created.

The study shows that the impacts of neighborhood services are not only important for health and environment, but also have importance in economic measures like increase in property values, increase in rent, decease transportations cost when a property is located near the road, and other locational expenditures. It has been analyzed that green spaces have a positive impact on sales and rental property values. A house/property near the road will have more value as well as more rent compared to others located away from the road. However, one of the most significant aspects found in the biological aspect of green spaces is whether they are present in core urban or peri-urban area, and they stabilize and attract the tourists around the city. But due to increasing rate of land requirements for settlement in urban areas, this has caused a decrease in natural land by conversion of land into built-up area. Due to the man-made revolution and development, one side gains benefits by making life easiest, but on the other side of the corner also has negative impacts on the natural environment, resources, and atmosphere as observed in study areas. Because of the higher rate of change of urban population due to rural to urban migration, the agricultural land is converted into urban environment. However, the study found that there is a strong need to establish more parks in all areas of the cities where the population exists. The local government should adopt a proper plan for land allocation for hospitals, market areas, and settlement. The study concluded that the rapid rate of migration and unregulated property market values and higher rate of demand increase the property as well as the land values.

The assessment of the impact of neighborhood's property on values is still vague. Therefore, a new agenda by comparative studies that use other housing and property-related services or parameters will be done on such issues; fruitful results could be obtained in the form of profit gain and urban development plans for residential settlements. These other services may include the security services, economic opportunity in the area, social security, energy supply, and provisions of municipal services in the area or region. These services are closely related to UN goal 11 for affordable housing and services supply.

The easy access to neighborhoods services increased the economic values. The results of this study will be the solid evidence based upon actual values and show significant effects on economic values, which supported the allocation of settlements and city development. Every type of neighborhoods amenity ensures the quality of life and increases the property values. However, some of the negatives points also noted like a property near the road will provide the proximity, but it will also create noise pollution. The traffic on the road affects the public privacy among the residents located near the major road. Therefore, effective urban planning is needed to overcome these issues for development and management.

Finally, the research suggested that property dealers and local government implement efficient land management and urban fringe development policies to land allocation planning for land available for use. The open land available adjacent to the roadside must be used for marketing areas for proximity effect for marketing and business. Residential area for new settlements should be separated from commercial area. These values could be used and adapted to increase awareness among the population, particularly property buyers and business community. These considerations and recommendations will improve policy decision and planning. Urban migration and sprawl should be monitor and controlled to create sustainability in urban infrastructure and management. This study provided a well-documented and research-based guideline and analysis tool for enhancing baseline information for urban land-use planning to increase the standards of residential land as well as its planning in urban areas and towns.

References

Aram, F., Solgi, E., & Holden, G. (2019). The role of green spaces in increasing social interactions in neighborhoods with periodic markets. *Habitat International, 84*, 24–32. https://doi.org/10.1016/j.habitatint.2018.12.004

Aziz, A., Anwar, M. M., & Dawood, M. (2021). The impact of neighborhood services on land values: An estimation through the hedonic pricing model. *GeoJournal, 86*(4), 1915–1925. https://doi.org/10.1007/s10708-019-10127-w

Brereton, F., Clinch, P., & Susana. (2008). Happiness, geography and the environment. *Ecological Economics, 65*(2), 386–396. https://doi.org/10.1016/j.ecolecon.2007.07.008

Chu, H., Venevsky, S., Wu, C., & Wang, M. (2019). NDVI-based vegetation dynamics and its response to climate changes at Amur-Heilongjiang River basin from 1982 to 2015. *Science of the Total Environment, 650*, 2051–2062. https://doi.org/10.1016/j.scitotenv.2018.09.115

Corrado, C., Hulten, C., & Sichel, D. (2005). Measuring capital in the new economy. In C. Corrado,

J. Haltiwanger, & E. D. Sichel (Eds.), *Measuring capital in the new economy* (pp. 11–46). University of Chicago Press. Retrieved from http://www.nber.org/books/corr05-1

Cui, N., Gu, H., Shen, T., & Feng, C. (2018). The impact of micro-level influencing factors on home value: A housing price-rent comparison. *Sustainability, 10*(12), 4343.

Diao, M. (2007). *Measuring the impact of urban form on property values—A hedonic price analysis*. Massachusetts Institute of Technology. Retrieved from https://ocw.mit.edu/courses/urban-studies-and-planning/11-434j-advanced-topics-in-real-estate-finance-spring-2007/projects/diao.pdf

Dunnett, N., Swanwick, C., & Woolley, H. (2002). *Improving urban parks, play areas and open spaces*. U. S. Department of Landscape.

Farinha-Marques, P., Fernandes, C., Gaio, A. R., Costa, J. P. D., & Guilherme, F. (2016). A sampling methodology to facilitate biodiversity assessment in public green spaces. *Urban Forestry & Urban Greening, 20*, 218–226. https://doi.org/10.1016/j.ufug.2016.09.004

Ferreira, S., Liu, H., & Brewer, B. (2018). The housing market impacts of wastewater injection induced seismicity risk. *Journal of Environmental Economics and Management, 92*, 251–269. https://doi.org/10.1016/j.jeem.2018.08.006

Georgi, J., & Dimitriou, D. (2010). The contribution of urban green spaces to the improvement of environment in cities: Case study of Chania, Greece. *Building and Environment, 45*(6), 1401–1414. https://doi.org/10.1016/j.buildenv.2009.12.003

Ham, C. (2011). *Using the hedonic property method to value federal lands proximate to urban areas*. Retrieved from https://dspace.library.colostate.edu/bitstream/handle/10217/46375/Ham_colostate_0053A_10617.pdf?sequence=1

Hanauer, M. M., & Reid, J. (2017). Valuing urban open space using the travel-cost method and the implications of measurement error. *Journal of Environmental Management, 198*, 50–65. https://doi.org/10.1016/j.jenvman.2017.05.005

Heinze, J. (2011). *Benefits of green space—Recent research*. Retrieved from http://www.ehrf.info/wp-content/uploads/2011/09/BenefitsofGreenSpace.pdf

Kabisch, N., Qureshi, S., & Haase, D. (2015). Human–environment interactions in urban green spaces—A systematic review of contemporary issues and prospects for future research. *Environmental Impact Assessment Review, 50*, 25–34. https://doi.org/10.1016/j.eiar.2014.08.007

Kausar, A., Zubair, S., Sohail, H., Anwar, M. M., Aziz, A., Vambol, S., Vambol, V., Khan, N. A., Poteriaiko, S., Tyshchenko, V., Murasov, R., Ejaz, F., & Khan, O. I. (2024). Evaluating the challenges and impacts of mixed-use neighborhoods on urban planning: An empirical study of a megacity, Karachi, Pakistan. *Discover Sustainability, 5*(1), 24. https://doi.org/10.1007/s43621-024-00195-5

Lafortezzaab, R., Carrusc, G., Sanesia, G., & Daviesd, C. (2009). Benefits and well-being perceived by people visiting green spaces in periods of heat stress. *Urban Forestry & Urban Greening, 8*(2), 97–108. https://doi.org/10.1016/j.ufug.2009.02.003

Luximon, Y., Kwong, H. Y., & Tai, Y. Y. (2015). User preferences of Urban Park seating pattern in Hong Kong. *Procedia Manufacturing, 3*, 4273–4278. https://doi.org/10.1016/j.promfg.2015.07.415

Mayora, K., Lyonsa, S., Duffya, D., Tola, R., & Juli, S. (2009). *Hedonic property value modeling of water quality, lake proximity, and spatial dependence in Central Florida [a dissertation, University of Central Florida Orlando, Florida]*. https://www.esri.ie/pubs/WP331.pdf.

Metz, N. (2015). Effect of distance to schooling on home prices. *Review of Regional Studies, 45*, 151–171. https://doi.org/10.52324/001c.8060

Millen, D. P. M. (2003). The number of subcenters in large urban areas. *Journal of Urban Economics, 53*(3), 321–338. https://doi.org/10.1016/S0094-1190(03)00026-3

Nha, D. V. (2017). Chapter 7 - The role of land-use planning on socioeconomic development in Mai Chau District, Vietnam. In M. Van Thanh, T. Duc Vien, S. J. Leisz, & G. P. Shivakoti (Eds.), *Redefining diversity & dynamics of natural resources management in Asia* (Vol. 2, pp. 87–111). Elsevier. https://doi.org/10.1016/B978-0-12-805453-6.00007-3

Panduro, K. E., & Veie, K. L. (2013). Classification and valuation of urban green spaces—A hedonic house price valuation. *Landscape and Urban Planning, 120*, 119–128. https://doi.org/10.1016/j.landurbplan.2013.08.009

Peschardt, K. K., Schipperijn, J., & Stigsdotter, U. K. (2012). Use of small public urban green spaces (SPUGS). *Urban Forestry & Urban Greening, 11*(3), 235–244. https://doi.org/10.1016/j.ufug.2012.04.002

Peschardt, K. K., & Stigsdotter, U. K. (2013). Associations between park characteristics and perceived restorativeness of small public urban green spaces. *Landscape and Urban Planning, 112*, 26–39. https://doi.org/10.1016/j.landurbplan.2012.12.013

Prunty, P. (2016). *An investigation into the effectiveness of hedonic features in regression models for domestic rental prices* [dissertations, Dublin Institute of Technology]. Retrieved from https://arrow.dit.ie/cgi/viewcontent.cgi?referer=https://www.google.com.pk/&httpsredir=1&article=1090&context=scschcomdis.

Sandera, & Polasky, S. (2009). The value of views and open space: Estimates from a hedonic pricing model for Ramsey County, Minnesota, USA. *Land Use Policy, 26*(3), 837–845. https://doi.org/10.1016/j.landusepol.2008.10.009

Sangwan, A., Saraswat, A., Kumar, N., Pipralia, S., & Kumar, A. (2022). *Urban green spaces prospects and retrospect's*. IntechOpen. https://doi.org/10.5772/intechopen.102857

Schreinemachers, P., & Berger, T. (2006). Land use decisions in developing countries and their representation in multi-agent systems. *Journal of Land Use Science, 1*, 29–44. https://doi.org/10.1080/17474230600605202

Wang, Q., Xiong, K., Zhou, J., Xiao, H., & Song, S. (2023). Impact of land use and land cover change on the landscape pattern and service value of the village ecosystem in the karst desertification control [original research]. *Frontiers in Environmental Science, 11*, 1020331. https://doi.org/10.3389/fenvs.2023.1020331

WHO. (2016). *Urban green spaces and health, Denmark*. Retrieved from http://www.euro.who.int/_data/assets/pdf_file/0005/321971/Urban-green-spaces-and-health-review-evidence.pdf?ua=1

WHO. (2017). *Urban green spaces*. World Health Organization Regional Office for Europe. Retrieved from https://www.dataplan.info/img_upload/5c84ed46aa0abfec4ac40610dde11285/who-urban-spaces.pdf

Wu, J. (2006). Environmental amenities, urban sprawl, and community characteristics. *Journal of Environmental Economics and Management, 52*(2), 527–547. https://doi.org/10.1016/j.jeem.2006.03.003

Zhang, H., Chen, B., Sun, Z., & Bao, Z. (2013). Landscape perception and recreation needs in urban green space in Fuyang, Hangzhou, China. *Urban Forestry & Urban Greening, 12*(1), 44–52. https://doi.org/10.1016/j.ufug.2012.11.001

Improving Urban Quality Through Ecological Parameters

Asad Aziz and Muhammad Mushahid Anwar

Abstract

Ecological parameters are the measurable variables that are determinants of an ecosystem. In recent years, negative impacts of urban activities show increasing signs of environmental problems in an ecosystem that degrade the urban life quality. To understand and overcome this critical situation in urban areas, it is necessary to study the interaction between living organisms and the urban environment. Therefore, for the establishment of a sustainable built environment for improving life quality ecological planning is an effective requirement in urban areas. This chapter defines the conceptual framework for developing balanced urban ecosystems through scientific methods and ecological approaches for planning by bibliometric analysis. The methods involved the development of a conceptual framework that creates a logical relationship between citizens, municipalities, and understand the relationship between the physical aspects and humans through the management of natural resources to improve urban life quality. The main outcome of the research is to develop a methodological framework and model for the assessment of the relationship between urban resources, urban life, and ecological parameters. The understanding of these parameters will improve the effectiveness of spatial planning practices that contribute to improve the urban life, through which effective urban planning can be accomplished.

A. Aziz (✉) · M. M. Anwar
Department of Geography, University of Gujrat, Gujrat, Punjab, Pakistan
e-mail: asadaziz@uog.edu.pk; mushahid.anwar@uog.edu.pk

Keywords

Analysis · Ecological · Natural · Parameters · Planning · Sustainable · Life · Urban life quality · Resources · Urban environment

Introduction

Urban life quality can be treated as a generic indicator of the physical and socio-economic condition of the urban and built environment. It provides a sense of satisfaction to the local population by evaluating different urban, socio-economic, and environmental parameters (König et al., 2023). The analysis of urban life quality is key for urban planning that influences to build a smart city (Deng, Wang, et al., 2023). The concept of a smart city gained substantial attention in the framework of urban developing strategies. The ecological parameters focus on approaches that estimate quality objectives at urban, environmental, and structural levels (Garau & Pavan, 2017). Research revealed qualitative and quanti-

M. M. Anwar, A. Aziz (eds.), *Sustainable Landscape for Urban Resilience*, Sustainable Development Goals Series, https://doi.org/10.1007/978-3-031-99836-2_6

tative indicators that can be useful for improving urban environmental quality. In the literature, a sustainable city is a city with a balance environmental focus, smart technologies, urban infrastructures, urban metabolism, water, energy, and waste management (Matsuda et al., 2023). To ensure livable conditions, urbanization loads have a very deeper consideration of the concept of a smart city, and urban planners, urban managers, and ecologists are discovering smoother methods to manage them. Sustainability is the capacity to systematize processes and functionality of the city in all sectors by stimulating economic growth, protecting the environment, and improving the urban life quality (Garau & Pavan, 2017). The concept of a sustainable smart city is often considered a philosophical aspect in the field of urban planning (Liang et al., 2020). Urban areas recently became conscious of the concept of smart cities by generating data in every field of management as well as strategies to understand future scenarios (Anees et al., 2013). They find the balance between the natural environment and resources through ecological parameters for good life quality and try to make the future bright (Narulita et al., 2016).

A generic indicator for improving the accuracy of urban environmental quality (UEQ) is the provision of sustainable ecological parameters (Khalil, 2012). This indicator represents the socio-economic, environmental, and social conditions of a built environment within an ecosystem (Hu et al., 2020). The value of these indicators expresses the sense of satisfaction of the local population (Faisal & Shaker, 2017). Ecosystem services in an urban environment are the basic tool to overcome the issues and challenges faced today by the urban population (Gavrilidis et al., 2016). These basic tools are green infrastructure, environmental justice, surrounding facilities, and social support services.

Dramatic economic development and large-scale urbanization along with excessive consumption of natural resources create serious challenges for urban life quality (Bian et al., 2020). These natural assets become increasingly scarce due to environmental protection and a lack of resources (Faisal & Shaker, 2017). So, the assessment of these parameters is the need of the day. Several ecological processes and parameters play a vital role in the ecosystem for the long term, therefore researchers face the challenges for ecological studies in the design of the study, nature of data, and choice of data (Verniest & Greulich, 2019). This chapter defines the conceptual framework for developing balanced urban ecosystems through scientific methods and ecological approaches for planning. The methods involved the development of a conceptual framework that creates a logical relationship between citizens and municipalities, and understanding the relationship between the physical aspects and humans through the management of natural resources to improve urban life quality. The main outcome of the research is to develop a methodological framework and model for the assessment of the relation between urban resources, urban life, and ecological parameters.

The first phase of this research is to formulate a methodology that creates a logical relationship between citizens and municipalities and understand the relationship between the physical aspects and humans through the management of natural resources to improve urban life quality. The second step of this framework is the study of the relationship between the physical aspects and humans through the management of natural resources to improve urban life quality. This relationship will provide support to promote the solution for sustainability challenges, including social, environmental, economic, and global levels that lead to an enhanced quality of life. This is the reason scholars around the world started to deliberate the term "sustainable smart cities."

As a notion urban quality of life has been discussed in various studies (Kijewski, 2020; Louis et al., 2023; Meryk et al., 2023; Serag El Din et al., 2013). It is a study to respond to many complications and challenges facing the world today. The purpose of this study involved analysis and provision of urban quality of life in precise terms quality of life, quality, and sustainable urban planning. The results of the study demonstrate the notion of sustainable development and also try to understand the relationship between the physical environment and human society.

Finally, the research concludes the principles for present urban planning for future urban quality. These principles will provide guidelines for the design process of urban life and policy making.

Review Literature

Worldwide planners and researchers have emphasized the need of planning for a better life (Kwon et al., 2023). Highly interrelated and extremely dynamic systems can be seen in the cities. The relationship between nature and people is a key nexus of environmental impacts (Elmqvist et al., 2015). Thus, this interrelated and dynamic system creates spatial heterogeneity in the urban system (Estévez-Mauriz et al., 2017). Urban landscape as an indicator of the quality of life (Estévez-Mauriz et al., 2017). Rapid expansion and the current projection of the world population offer fundamental opportunities for health, basic units of life, and social justice, but on the other hand also create challenges (Elmqvist et al., 2015). Therefore, there is a strong need for analysis to overcome these issues and the causes which create these challenges (Plieninger et al., 2015).

Rapid urbanization cause the accelerated degradation of the urban environment (Liu et al., 2017). In recent years, negative impacts of urban activities show increasing signs of environmental problems in an ecosystem that degrade the urban life quality (Shin & Jung, 2020). These signs are pressure on cities which is assessed through the notion of ecological footprint (Wackernagel & Kitzes, 2008). The major concerns for cities that cause this degradation are climate change, degradation along with depletion of natural resources, and development pressure on green areas. This imbalance of nature threatens life around the globe (Sampaio et al., 2022). To understand and overcome this critical situation in urban areas it is very necessary to study the interaction between living organisms and the urban environment. Thus, urban ecology is fairly important for sustainable urban planning (Niemelä, 1999). A very important function of urban life quality is to establish a sustainable environment for urban dwellers, communities, and natural systems (Nautiyal & Goel, 2021). Therefore, for the establishment of a sustainable built environment for improving life quality ecological planning is an effective requirement in urban areas.

Urban management is the basic element for ensuring the quality of life, urban management means efficient strategies and planning which ensure the environmental quality in urban areas (Astiaso Garcia, 2017). A key component of sustainable development is the provision of quality of life to urban dwellers (Stessens et al., 2020) (Duran et al., 2015). The primary source of life quality in contact with nature. Quality of life is the first aim and major concern for sustainable development that is evaluated through various parameters, these parameters are much related to the consumption of environmental services, environmental quality, and environmental behavior (Streimikiene, 2015).

Urbanization is a great modifier for ecosystem service (Malmqvist et al., 2009). It is a global multidimensional process that manifests the rapid changing of the population and local landscape of an area (Elmqvist et al., 2008). Today, urbanization is viewed as more geographically ubiquitous than any other activity (Deng, Yang, & Wang, 2023), which alters the flora and fauna and changes the ecosystem cycle as a part of ecological parameters in the given space (Elmqvist et al., 2008). So, there is a fundamental need to analyze and study these parameters for improvement in human life and maintain the natural cycle of the ecosystem.

Methodology

Background

This paper defines the conceptual framework for developing a balanced urban ecosystem through scientific methods and ecological approaches for planning and development through a comprehensive instigation of past data. The methods involved the development of a conceptual framework that creates a logical relationship between citizens and municipalities, and understanding the relationship between the physical aspects and humans through the management of natural resources to improve urban life quality. The methods used in the analysis synthesize the conceptual and theoretical dimensions for the determination of urban life quality.

Technique and Methods

These ecological approaches are categories and indicators that are further divided into subindicators. In this study, these ecological parameters/approaches as indicators are the ***use and function*** of ecosystem services (IQ USE), ***health and well-being*** (IQ H&W), ***appearance*** (IQ APP), management (IQ MAN), environment (IQ ENV), and ***safety and security*** (IQ SEC).

The provision of these facilities and services to all individuals in a given environment will ensure the quality of life in an area (Chang et al., 2017). The sum of all of these services is equal to the quality of life that is elaborated in the following equation (adapted from Bian et al., 2020):

$$\mathrm{ULQ} = \mathrm{IQ\,USE} + \mathrm{IQ\,H\,\&\,W} + \mathrm{IQ\,APP} + \mathrm{IQ\,MAN} + \mathrm{IQ\,ENV} + \mathrm{IQ\,SEC}$$

where *ULQ* urban life quality, *IQ USE* function of ecosystem services, *IQ H&W* health and well-being, *IQ APP* appearance, *IQ MAN* management, *IQ ENV* environment, and *IQ SEC* safety and security. Table 1 further explains these categories and indicators with their further division.

In the above technique, different components are considered in the study that provide and define the life quality in urban areas. These components in the form of categories are value-added services, including personal development, security, health, sustainable environment, natural resources, ecosystem/environmental services, comfort, personal satisfaction, and feelings regarding their living and lifestyle in a given environment. Indeed, life depends on the distribution and presence of activities and space in cities. Ultimately, the presence and equal distributions of all the abovementioned services provide a quality of life to the local inhabitant. The greater the use of services, more the easiness in life and better the quality; this function is categorized in use and function.

Finally, security and security refer to attitudes, perceptions of personal safety and security, and social justice. The attitude is related to the risk involved in a traffic accident, victim of an offense or criminal case, or any other social violence, or threats. Being a human, we are always interested in ensuring our safety and the safety. As an ecological parameter, it is a fundamental need nowadays.

The indicators described above are used as tools in planning for improving the quality of life in urban regions in qualitative as well as quantitative dimensions (McCrea, 2007). These categories and indicators are the products and will have a sum, and their sum aggregate gives the value that is called urban life quality (ULQ). In this study, the urban life quality (ULQ) is based upon a general layout that may call a structural procedure from which categories, indicators, and subindicators are created for analysis to get the results.

A complete master plan was prepared and rating system was formulated with a questionnaire, and an in situ neighborhood analysis was conducted. After data conduction, the theoretical analysis was done for results. To create a rating for each neighborhood census area, the study identifies aspects that present the greatest problems for each area; after identification, the investigation results. Finally, these results were presented to planners for improving the neighborhood's environmental quality.

Results

The results of the study are qualitative. The first outcome of this study is the identification of analysis of previously listed criteria for the assessment of urban life quality. The study shows that improving life quality is a measurement of factors required to bring positive changes in a given environment and fulfill the need of local dwellers. The results of the study attempt to measure the quality of urban life in the city. Therefore, to facilitate the use of standards and indicators will

Table 1 Categorization of urban sustainability indicators and their subdivisions

	Factors driving land uses evidence	
Categories	Physical indicators	
Use and function	Accessibility	Traffic accessibility Pedestrian accessibility and housing structure Sustainable walkability Accessibility for disables Cyclability
	Flexibility and functionality	Service for the people with disability Multifunctional and sustainable urban equipment
	Minimum service provided	Availability of services and equipment Availability of waste container
	Social indicators	
Health and well-being	Emotional well-being quality of street lighting Environment maintenance	Presence of green areas Attractive living places Easy mobility services
	Quality of life	Urban traffic noise pollution Minimum air pollution Housing livability Sustainability of public spaces Presence of public spaces Sustainable service and healthcare facilities
	Social well-being	Open spaces for children and gaming Economic opportunities Perception of security
Appearance	Environmental properties	Quality of urban landscape Green maintenance
	Built-up characteristics	Urban design maintenance Quality of housing and aesthetic characteristics
Management	Efficiency of primary services	Waste management The efficiency of urban maintenance services
Environment	Soil pollution	Protection of soil fertility Respect for land use and landscape Cleaning of streets and public areas Waste management and disposal point availability Access to clean and safe water
Safety and security	Security system Smart crime prevention	Lighting and security system Social security level
	Risk of natural disaster	Probability of natural disaster

Adapted from Bian et al. (2020)

be used for analyzing the performance of a given urban area setting priorities and setting targets for change for further improvement.

The results provide a methodology in relation to two different urban realities, including the framework for studying the relationship between the physical environment and humans and secondly on the basis of this relation set the dimension for the future. With regards to the qualitative section of the results, the evaluation was made several times, which included the general observation of places, then assessment of people's life quality through interviews and questionnaires, and finally by reviewing the available data in the form of reports, public documents, and new papers/reports. The available data related to health and well-being, neighborhood amenities, social values, traffic noise, presence of green spaces, justice, safety and security, urban landscape appearance, and environment as all of these indicators were considered in the methodology section. The complete evaluation of all of these indicators gives us the values and magnitude of the urban life quality index.

Despite various studies and projects on urban studies related to ecological parameters, urban sprawl and urban renewal show significant improvements in cities in recent years (Karakayaci, 2016). The results of the study express that none of the cities yet be described are an example of a smart sustainable city and urban best practices. Results show that problems related to justice, social integration, services, facilities, lack of a safe environment, and equal access to affordable housing affect the quality of life and disturbed the magnetism of ecological parameters. The advantages of such studies are linked with policymaking processes and good governance that also attract the attention of concerned authorities.

Discussion

The urban (Barr & McNamara, 2022) life quality (ULQ) is considered a fundamental element of an ecosystem with different environmental services in built-up areas and their efficiency. This includes the urban landscape and green infrastructure, accessibility to services, social justice, security, economic opportunity, good governance, and the quality of the house (Cai et al., 2021). A city has a good appearance with wide road and street networking, green spaces, and infrastructure, a well-structured buildup area, and a local cultural base landscape (Cai et al., 2021). The category health is related to well-being and health with the combined effects of social, physical, biological, and physiological well-being of people. These aspects are related to personal, contextual, and environmental characteristics of health and livability. The category management and environment are related to municipal series their management and availability in an area. Good management of the environment is key to ensure the quality (Oyebode et al., 2023).

The well-being of a society is a measure of its success, so people will move and live there where society has all of these measures (van Hoof et al., 2019). That is the reason it is necessary to understand the nature of local people/residents, happiness, and suggest policies and direction for improvement. Local population engagement plays a vital role in this aspect. Citizen involvement, responsive policies, and public administration are fundamental tools for improving services and finding solutions (ten Bruggencate et al., 2019). Participation of local people is value-added, which increases the efficiency and effectiveness of urban policies that improve and ensure the quality of living standards and employment opportunities.

Quality of life is an umbrella term; it is a complicated cycle of an ecosystem, including the social, physical, environmental, and ecological domains of humans around the globe (Magiera & Pac, 2022). For example, having a hobby in life is good for social health. A park or green infrastructure near the settlement ensures the equality of physical health (Kruk et al., 2018). A sense of security in a human society is an example of social health, and a strong association between abiotic and their maximum role-play for human well-being defines the quality of ecological life in a society. This will ensure and strengthen the social–ecological association in an environment. However, there are some challenges and hurdles in the least developing and developed nations over the globe (ten Bruggencate et al., 2019).

Conclusion

For city planners, quality of life is a strategic issue in urban areas. To move toward sustainability and smartness, cities are excellent tools for the implementation of urban planning policies. Ecological parameters provide forceful assessment tools that encompass social, environmental, and urban planning as a subjective estimation. By achieving the objectives of the study, the results concluded the evaluation of urban quality levels that contribute to further planning for the cities and play a significant role in achieving the smartness of urban areas around the world. However, the quality of life depends on different parameters, and these parameters vary from location to location and area to area with different local, regional, and global scales. The significance of the applied method is to embed the possibility of adaptation that is very common and flexible according to the real situation and can be adjustable for practical implementation in developed and developing nations. The results of the study allow the city planner, managers, urban ecologists, and public authorities to promote integrated urban governance through effective governance actions and monitoring. This is the main theme of contemporary sustainable development in the field of urban planning and sustainable development.

Quality of life is a global term with different definitions and approaches that encompass multi-structural and multidimensional parameters of bio life and their ecosystem at a given time and place. Today, the rate of human dependency on nature and natural resources is accelerating due to changes in technological and regional complex approaches covering different aspects of human life. There is a dire need to create a strong association of human needs with the environment, and there should be a balanced approach with nature and humans for existence in the world.

References

Anees, M. M., Mohamed, H. E., & Abdel Razek, M. E. (2013). Evaluation of change management efficiency of construction contractors. *HBRC Journal, 9*(1), 77–85. https://doi.org/10.1016/j.hbrcj.2013.02.005

Astiaso Garcia, D. (2017). Green areas management and bioengineering techniques for improving urban ecological sustainability. *Sustainable Cities and Society, 30*, 108–117. https://doi.org/10.1016/j.scs.2017.01.008

Barr, M. L., & McNamara, J. (2022). Community-based participatory research: Partnering with college students to develop a tailored, wellness-focused intervention for university campuses. *International Journal of Environmental Research and Public Health, 19*(23), 16331. https://doi.org/10.3390/ijerph192316331

Bian, J., Ren, H., & Liu, P. (2020). Evaluation of urban ecological Well-being performance in China: A case study of 30 provincial capital cities. *Journal of Cleaner Production, 254*, 120109. https://doi.org/10.1016/j.jclepro.2020.120109

Cai, T., Verze, P., & Bjerklund Johansen, T. E. (2021). The quality of life definition: Where are we going? *Uro, 1*(1), 14–22. https://doi.org/10.3390/uro1010003

Chang, K. A.-O., Wong, F. K. Y., Chan, K. L., Wong, F., Ho, H. A.-O., Wong, M. S., & Yang, L. (2017). The impact of the environment on the quality of life and the mediating effects of sleep and stress. *International Journal of Environmental Research and Public Health, 17*(22), 8529.

Deng, H., Yang, J., & Wang, P. (2023). Study on coupling coordination relationship between urban development intensity and water environment carrying capacity of Chengdu–Chongqing economic circle. *Sustainability, 15*(9), 7111. https://doi.org/10.3390/su15097111

Deng, S.-Y., Wang, Y.-Z., Peng, M.-M., Zhang, T.-M., Li, M., Luo, W., & Ran, M.-S. (2023). Quality of life among family caregivers of people with schizophrenia in rural China. *Quality of Life Research, 32*(6), 1759–1769. https://doi.org/10.1007/s11136-023-03349-6

Duran, D., Gogan, L., Artene, A., & Duran, V. (2015). The components of sustainable development - A possible approach. *Procedia Economics and Finance, 26*, 806–811. https://doi.org/10.1016/S2212-5671(15)00849-7

Elmqvist, T., Alfsen, C., & Colding, J. (2008). Urban Systems. In S. E. Jørgensen & B. D. Fath (Eds.), *Encyclopedia of ecology* (pp. 3665–3672). Academic Press.

Elmqvist, T., Setälä, H., Handel, S. N., van der Ploeg, S., Aronson, J., Blignaut, J. N., & de Groot, R. (2015). Benefits of restoring ecosystem services in urban areas. *Current Opinion in Environmental Sustainability, 14*, 101–108. https://doi.org/10.1016/j.cosust.2015.05.001

Estévez-Mauriz, L., Fonseca, J. A., Forgaci, C., & Björling, N. (2017). The livability of spaces: Performance and/or resilience? Reflections on the effects of spatial heterogeneity in transport and energy systems and the implications on urban environmental quality. *International Journal of Sustainable Built Environment, 6*(1), 1–8. https://doi.org/10.1016/j.ijsbe.2016.10.001

Faisal, K., & Shaker, A. (2017). Improving the accuracy of urban environmental quality assessment using geographically-weighted regression techniques. *Sensors, 17*(3), 528.

Garau, C., & Pavan, V. M. (2017). Evaluating urban quality: Indicators and assessment tools for smart sustainable cities. *Sustainability, 10*(3), 18. https://doi.org/10.3390/su10030575

Gavrilidis, A. A., Ciocănea, C. M., Niţă, M. R., Onose, D. A., & Năstase, I. I. (2016). Urban landscape quality index—Planning tool for evaluating urban landscapes and improving the quality of life. *Procedia Environmental Sciences, 32*, 155–167. https://doi.org/10.1016/j.proenv.2016.03.020

Hu, D., Meng, Q., Zhang, L., & Zhang, Y. (2020). Spatial quantitative analysis of the potential driving factors of land surface temperature in different "centers" of polycentric cities: A case study in Tianjin, China. *Science of the Total Environment, 706*, 135244. https://doi.org/10.1016/j.scitotenv.2019.135244

Karakayaci, Z. (2016). The concept of urban sprawl and its causes. *Journal of International Social Research, 9*, 815–815. https://doi.org/10.17719/jisr.20164520658

Khalil, H. A. E. E. (2012). Enhancing quality of life through strategic urban planning. *Sustainable Cities and Society, 5*, 77–86. https://doi.org/10.1016/j.scs.2012.06.002

Kijewski, S. (2020). Life satisfaction sixty years after world war II: The lasting impact of war across generations. *Applied Research in Quality of Life, 15*(5), 1253–1284. https://doi.org/10.1007/s11482-019-09726-z

König, H.-H., Neumann-Böhme, S., Sabat, I., Schreyögg, J., Torbica, A., van Exel, J., & Hajek, A. (2023). Health-related quality of life in seven European countries throughout the course of the COVID-19 pandemic: Evidence from the European Covid survey (ECOS). *Quality of Life Research, 32*(6), 1631–1644. https://doi.org/10.1007/s11136-022-03334-5

Kruk, M. E., Gage, A. D., Arsenault, C., Jordan, K., Leslie, H. H., Roder-DeWan, S., & Pate, M. (2018). High-quality health systems in the sustainable development goals era: Time for a revolution. *The Lancet Global Health, 6*(11), e1196–e1252.

Kwon, J.-Y., Kopec, J., Sutherland, J. M., Lambert, L. K., Anis, A. H., & Sawatzky, R. (2023). Patient-reported mental health and well-being trajectories in oncology patients during radiation therapy: An exploratory retrospective cohort analysis using the Ontario cancer registry. *Quality of Life Research*. https://doi.org/10.1007/s11136-023-03430-0

Liang, Z., Wu, S., Wang, Y., Wei, F., Huang, J., Shen, J., & Li, S. (2020). The relationship between urban form and heat Island intensity along the urban development gradients. *Science of the Total Environment, 708*, 135011. https://doi.org/10.1016/j.scitotenv.2019.135011

Liu, Y., Yue, W., Fan, P., Zhang, Z., & Huang, J. (2017). Assessing the urban environmental quality of mountainous cities: A case study in Chongqing, China. *Ecological Indicators, 81*, 132–145. https://doi.org/10.1016/j.ecolind.2017.05.048

Louis, G., Pétré, B., Schleich, F., Zahrei, H. N., Donneau, A.-F., Henket, M., & Louis, R. (2023). Predictors of change in asthma-related quality of life: A longitudinal real-life study in adult asthmatics. *Quality of Life Research, 32*(5), 1507–1520. https://doi.org/10.1007/s11136-022-03339-0

Magiera, A., & Pac, A. (2022). Determinants of quality of life among adolescents in the Malopolska region, Poland. *International Journal of Environmental Research and Public Health, 19*(14), 8616. https://doi.org/10.3390/ijerph19148616

Malmqvist, B., Timm, H., Muotka, T., & Nilsson, C. (2009). Chapter 8 - The Fennoscandian shield. In K. Tockner, U. Uehlinger, & C. T. Robinson (Eds.), *Rivers of Europe* (pp. 297–335). Academic Press.

Matsuda, A., Yamaoka, K., Kunitoh, H., Seto, T., Tsuboi, M., Ohira, T., & Kubota, K. (2023). Quality of life with docetaxel plus cisplatin versus paclitaxel plus carboplatin in patients with completely resected non-small cell lung cancer: Quality of life analysis of TORG 0503. *Quality of Life Research*. https://doi.org/10.1007/s11136-023-03424-y

McCrea, R. (2007). Urban quality of life: Linking objective dimensions and subjective evaluations of the urban environment.

Meryk, A., Kropshofer, G., Hetzer, B., Riedl, D., Lehmann, J., Rumpold, G., & Crazzolara, R. (2023). Disagreement between mothers' and fathers' rating of health-related quality of life in children with cancer. *Quality of Life Research, 32*(6), 1683–1691. https://doi.org/10.1007/s11136-023-03341-0

Narulita, S., Zain, A. F. M., & Prasetyo, L. B. (2016). Geographic information system (GIS) application on urban forest development in Bandung City. *Procedia Environmental Sciences, 33*, 279–289. https://doi.org/10.1016/j.proenv.2016.03.079

Nautiyal, H., & Goel, V. (2021). Chapter 3 - Sustainability assessment: Metrics and methods. In J. Ren (Ed.), *Methods in sustainability science* (pp. 27–46). Elsevier.

Niemelä, J. (1999). Ecology and urban planning. *Biodiversity and Conservation, 8*(1), 119–131. https://doi.org/10.1023/A:1008817325994

Oyebode, O., Torres-Sahli, M., Kapinga, D., Bruno-McClung, E., Willans, R., Shah, N., & Stewart-Brown, S. (2023). Swahili translation and validation of the Warwick Edinburgh mental wellbeing scale (WEMWBS) in adolescents and adults taking part in the girls' education challenge fund project in Tanzania. *Health and Quality of Life Outcomes, 21*(1), 43. https://doi.org/10.1186/s12955-023-02119-9

Plieninger, T., Bieling, C., Fagerholm, N., Byg, A., Hartel, T., Hurley, P., & Huntsinger, L. (2015). The role of cultural ecosystem services in landscape management and planning. *Current Opinion in Environmental Sustainability, 14*, 28–33. https://doi.org/10.1016/j.cosust.2015.02.006

Sampaio, F., Salgado, R., Antonini, M., Delmas, P., Oulevey Bachmann, A., Gilles, I., & Ortoleva Bucher, C. (2022). Workplace wellbeing and quality of life perceived by Portuguese nurses during the COVID-19 pandemic: The role of protective factors and stressors. *International Journal of Environmental Research and Public Health, 19*(21), 14231. https://doi.org/10.3390/ijerph192114231

Serag El Din, H., Shalaby, A., Farouh, H. E., & Elariane, S. A. (2013). Principles of urban quality of life for a neighborhood. *HBRC Journal, 9*(1), 86–92. https://doi.org/10.1016/j.hbrcj.2013.02.007

Shin, S., & Jung, K. (2020). Mediating role of community participation between physical environments, social relationships, social conflicts, and quality of life: Evidence from South Korea. *Applied Research in Quality of Life, 15*(5), 1433–1450. https://doi.org/10.1007/s11482-019-09747-8

Stessens, P., Canters, F., Huysmans, M., & Khan, A. Z. (2020). Urban green space qualities: An integrated approach towards GIS-based assessment reflecting user perception. *Land Use Policy, 91*, 104319. https://doi.org/10.1016/j.landusepol.2019.104319

Streimikiene, D. (2015). Environmental indicators for the assessment of quality of life. *Intellectual Economics, 9*(1), 67–79. https://doi.org/10.1016/j.intele.2015.10.001

ten Bruggencate, T., Luijkx, K. G., & Sturm, J. (2019). Friends or frenemies? The role of social technology in the lives of older people. *International Journal of Environmental Research and Public Health, 16*(24), 4969. https://doi.org/10.3390/ijerph16244969

van Hoof, J., Beneken Genaamd Kolmer, D. M., de Vlugt, E., & de Vries, S. I. (2019). Quality of life: The interplay between human behaviour, technology and the environment. *International Journal of Environmental Research and Public Health, 16*(24), 5106. https://doi.org/10.3390/ijerph16245106

Verniest, F., & Greulich, S. (2019). Methods for assessing the effects of environmental parameters on biological communities in long-term ecological studies - A literature review. *Ecological Modelling, 414*, 108732. https://doi.org/10.1016/j.ecolmodel.2019.108732

Wackernagel, M., & Kitzes, J. (2008). Ecological footprint. In S. E. Jørgensen & B. D. Fath (Eds.), *Encyclopedia of ecology* (pp. 1031–1037). Academic Press.

Footprints Indicators and Biodiversity Loss: A Review of Ecological Informatics

Asad Aziz and Muhammad Mushahid Anwar

Abstract

Ecological informatics is a multidisciplinary framework for designing sustainable modeling using intensive spatial data, ecological footprint, and computational technology. This study reviews the past literature to identify the ecological networks influencing land-use change and their impacts on natural habitat and biodiversity of study area. Habitat change and loss of biodiversity are the major challenges for sustainable development and UN goals, particularly in the least developing states. The phenomena of dependency of human on natural resources are accelerating in thickly populated cities and creating complicated ecological relationship with mixed morphological characteristics, which need to address urban resilience and earth biocapacity. This past research considered the physical, ecological, social, and environmental constraints as ecological networks to identified eco-nature of area to design ecological models by using ecological informatics for urban managers and government institutions for conservation strategies and their effectiveness at a different scale. These strategies will drive the principles to mitigate the negative impacts of land-use change on biodiversity. The urban managers, land-use planners, policymakers, land managers, and relevant stakeholders use these policies in decision-making for society's well-being and urban justice.

Keywords

Bio-capacity · Earth · Ecology · Land · Land use · Urban · Indicators · Population · Justic · Urban Managers

Urbanization and Land Use

Urban areas around the world are the hub of population and built-up infrastructure with varying land-use character. Land-use fragmentation due to rapid urbanization puts significant pressure on biodiversity and habitat change and creates issues for ecological capacity of cities. This land use is a sign of human dependency on earth and its resources. The unplanned land use caused very serious negative issues for urban spatial planning and carrying capacity of the earth (Adu-Gyamfi et al., 2021). This carrying capacity issues have multidomains, including change in niche, species extinction, habitat change, deforestation, increase in temperature, due to more emission of CO_2, land surface temperature, and change in weather pattern. This issue is more

A. Aziz (✉) · M. M. Anwar (✉)
Department of Geography, University of Gujrat, Gujrat, Punjab, Pakistan
e-mail: asadaziz@uog.edu.pk; mushahid.anwar@uog.edu.pk

M. M. Anwar, A. Aziz (eds.), *Sustainable Landscape for Urban Resilience*, Sustainable Development Goals Series, https://doi.org/10.1007/978-3-031-99836-2_7

dominant in the urban world and least developing countries. Over the years, the urban population has expanded very rapidly. According to the United Nations reports, approximately 55% of the population is living in cities and is expected to increase by 68% in 2050 (Anthony, 2023). This rapid rate of urbanization and population growth is a challenging task for urban planners, managers, and policymakers for sustainable development and urban justice, which need to be addressed for society well-being. Sustainable urbanization is focused on building cities with economically vibrant and also have a balance between the environmental, ecological, political, social, and economic needs of the urban dwellers (Ashrafzadeh et al., 2019). These cities reduce pollution, increase economic vitality, socially vibrant, affordable housing, green infrastructure, freshwater bodies, strong transportation networks, and environmental justice (Bai et al., 2022). These cities have a strong network of urban planning and resource management that is responsive to the needs of their residents, government, stakeholders, and urban managers.

Population, Ecology, and World

It is a matter of fact that much of the population around the world, including developing and developed countries, is moving toward cities to improve quality of life and economic opportunities (Batty, 2019). This phenomenon is more dominant in developing states, where rural areas lack the basic facilities. The urban areas, on the other hand, have limited capacity to cope with the demands of the locals at a time (Bauer & Hoye, 2014). These demands are associated with environmental, economic, ecological, political, or social needs. Environmental demands include the provision of fresh air, water, and green infrastructure; the ecological demands include shelter for living, oxygen for survival, land for agriculture, and a sustainable ecosystem in the environment for a better life quality. The economic demands include job opportunities, economic services, and social demands, including the sense of belonging or community engagement, social justice, municipal services, equity, and security. It is necessary that if a city is a sustainable city, then it will have all of these services in the long term. However, the growing demands of the population are challenging for cities as well as for humans and carrying capacity of the earth as a planet or home for men (Bennett, 2017).

Around the world, urbanization puts significant pressure on natural environmental and causes many problems. Some of these impacts include the habitat fragmentations, change of habitat, and increase in death rate of species (Chaudhary et al., 2015), threat to biodiversity, deforestation, distribution of nature and wildlife, regional and global warming of climate. It has been investigated that these problems affect the urban sustainability and ecological pattern of an area (Cook & Pinter-Wollman, 2019), which need to be addressed for sustainable approach.

Around the world, the cities consume 60–70% of energy in different sectors, for example, industry, transportation, infrastructure development, and telecommunications, and contribute 70% of total carbon released in the world (Crane et al., 2021; Dastjerdi & Nasrabadi, 2022). This huge amount of carbon ultimately causes global warming, an increase in temperature, climate change, and natural disasters, which makes the cities vulnerable and leads to economic, social, and human loss. Only 3% of the land area over the entire planet earth is covered by cities, but their contributions to natural disasters are significant in terms of climate change, global warming, ozone depletion, floods, and pollution, along with other environmental, social, economic and ecological disruptions (De Montis et al., 2017). In sum, these issues create social segregation, environmental injustice, inequality, economic instability, climate gentrification, and unsustainability (De Palma et al., 2021).

Urban areas around the world are the hub of population and built-up infrastructure with varying urban characteristics and population around the globe. The fragmentation due to rapid urbanization puts significant pressure on global as well as regional natural resources that creates serious issues for the ecological capacity of cities mostly in the developing world, that is, Asia, Africa, and

South Asia, and the same case has been observed in the capital of Pakistan. This unplanned urban expansion challenges the urban sustainability as proved by (Demaidi, 2023) by disturbing the existing infrastructure of a city. The unplanned land use by rapid population creates serious negative issues for urban planning in spatial context and carrying capacity of the earth disturbs the local, regional, and global environment, society, and economy (Adu-Gyamfi et al., 2021). The carrying capacity issues have multiple domains, including change in the niche, species extinction, habitat change, unplanned land use, deforestation, increase in temperature, due to more emission of CO_2, change in land surface temperature, and change in weather patterns, along with other economic, social, and ecological issues like scarcity of municipal services, lack the sense of belonging, lack of social security, high land and rent values, degradation of the local ecosystem, decrease in the groundwater table, its quality, and environmental justice (Díaz et al., 2019). These issues are more dominant in the urban world of least developing regions like South Asia, sub-Saharan, and some countries in the Middle East (Din et al., 2023).

It has been investigated that cities around the world face multidomain problems affecting the regional sustainability and ecological pattern of cities (Cook & Pinter-Wollman, 2019). The concept and strategies of the sustainable city and its concept are varied spatially and temporally, and this concept is gaining more and more importance among leaders, thinkers, and managers for making the planet the home of man (Dudley & Alexander, 2017). The fundamental goal for a sustainable city is the provision of services to its citizens with maximum livelihoods and equity (Elias, 2018). A sustainable city is a different concept in developed and developing states around the world, and every country is trying to attain the picture of a sustainable city even in developed nations (García-Vega & Newbold, 2020). Sustainability means the minimum resource depletion with maximum efficiency, for improvement in men's life without damaging the environment, social and economic stability, and equity (Gonthier et al., 2014).

In the world, all of the capital cities are developed gradually in every phase of ecological, economic, geographical, political, technological, and social race (González-Marín et al., 2023). This development is closely associated with sustainable land-use planning, infrastructure development, capital structure, good governance, technology and innovation in transportation, monitoring of mass transit, information technology and innovations, green economy, justice and peace, and security and energy sectors. Around the world, artificial intelligence (AI) has been used by several countries, even in developing nations (Guan et al., 2022). This includes the response to energy crises, economy, airports and transportation, research and development, education, industries, food and agriculture, and healthcare centers to ensure the services (Hansen et al., 2004; He et al., 2021).

For sustainable development for the region and world, it is necessary to tackle the ecological, social, and economic concerns affecting the sustainability for short and long term with multiple scales (Hosseini et al., 2019). It is thinking today for a better tomorrow in the near future (Houlden et al., 2021). The overexploitation of natural resources by the population to meet the demands is a challenging task for the government and relevant stakeholders to sustain the ecosystem.

Land-Use Dynamics

The past studies proved a very strong association between landscape networks, land-use cover and land-use change, and biodiversity, which create different ecological patterns of urban landforms and their networks. This study aims to provide an in-depth picture of these ecological networks by considering multiple ecological indicators taken as ecological informatics. This network of informatics is related to social, economic, physical, and environmental constraints of an area (Hsu et al., 2022). Among all of these, the environmental constraints are strongly related to green spaces and biodiversity, which can lead the fragmentations in landscapes and elimination of green patches as investigated by previous literature (Hu

& Li, 2012). The process of fragmentations is dynamic and changes over time due to natural events and human activities over the years (Ionuţ-Constantin & Eugen, 2016). Therefore, the assessment of this change, along with an understanding of these ecological dynamics of landscape, is needed.

It has been observed that the urban ecologist faces many challenges and issues during the designing of morphological connection of ecological networks due to limited resources and competing priorities, which may lead to unrealistic results (Jackson et al., 2013). However, the assessment of these ecological informatics can be rationalized through environmental input–output estimations. This estimation will give the picture of bioproductive capacity to sustain biodiversity and human population. This investigation can be done by assessing the lifestyle of local population to access their ecological footprints and used as ecological informatics (Hu & Li, 2012).

Various methods have been used for connectivity analysis between landscape and their connection with biodiversity and green areas (Javanmardi et al., 2023). This network includes gravity modeling, circuit theory, least-cost approach, and networker spaces (Kok et al., 2020). This study aims to apply ecological informatics to identify the relationship between land-use change and its impacts on biodiversity and link with remote sensing to get a spatial view of the area under examination. The study uses land-use matrix to quantify the changes in landscape and its connectivity with social, physical, and ecological parameters and identify the core area of biodiversity and potential corridors to evaluate the ecological informatics. The remote sensing data help us understand the landscape fragments and provide potential application for modeling the future pattern of land-use changes (Kondratyeva et al., 2020), which will drive the suggestion for the area under examination for the future.

Methods

Morphology of Ecological Indicators

At present, the complicated cycle of ecology sharpens the complex morphology link between biotic and abiotic components in an ecosystem (Kvasničková Stanislavská et al., 2020). This morphological character of ecosystem needs multidisciplinary intensive data set from different disciplines for making the decisions, analysis, policymaking, and disaster management. This cycle consists of different entities, including (organisms, genomes, communities, population, landscape, and ecosystem) (Leclère et al., 2020). Ecological informatics is a complex cycle consisting of different phases from plan to data management, its synthesis, and presentation (Lin et al., 2021).

Figure 1 shows the major nodes of ecological networks, which are further divided into subsections, defining the characteristics of the entire area in physical, social, and environmental parameters. Table 1 shows the identified ecological informatics taken into consideration for further analysis and as evidence from previous studies.

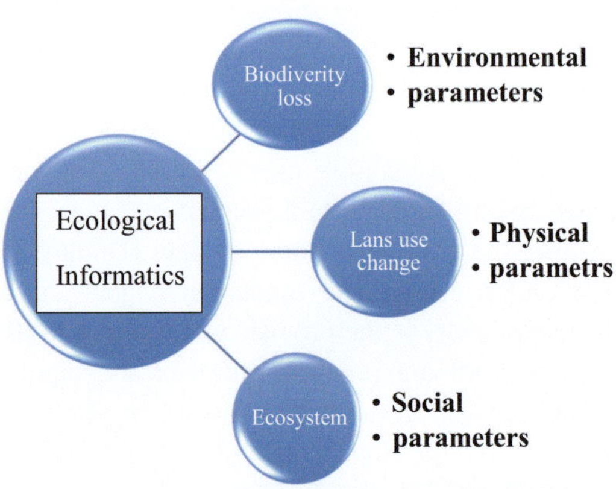

Fig. 1 Conceptual framework of ecological network

Table 1 Hierarchal modeling of ecological network

S. No.	Factors driving land uses	Evidence
Physical informatics		
1	Land-use diversity	Ionuţ-Constantin and Eugen (2016), Kok et al. (2020), Leclère et al. (2020), Loures et al. (2007), Malla and Bandh (2023), Mamun and An (2022), Marselle et al. (2021)
2	Urbanization and land-use intensity	Adu-Gyamfi et al. (2021), Chaudhary et al. (2015), Ionuţ-Constantin and Eugen (2016), Leclère et al. (2020), Lin et al. (2021), Malla and Bandh (2023), Martens et al. (2022)
3	Poor dwelling and housing infrastructure	Lin et al. (2021), Meng (2019), Meyer et al. (2022), Michener (2018), Moallemi et al. (2020)
4	Urban infrastructure and biodiversity loss	Molinos et al. (2018), Musa et al. (2023), Muscat et al. (2020)
5	Road networks and public pathways	Kvasničková Stanislavská et al. (2020), Nations (2015, 2023), Olanrewaju et al. (2015)
Social informatics		
1	Lack of public parks/green infrastructure	Park (2022), Pascual et al. (2013), Paulin (2019a, 2019b)
2	Low public health and well-being	Kvasničková Stanislavská et al. (2020), Muscat et al. (2020), Peters et al. (2013), Porter (2018), Ramirez-Yanez et al. (2018)
3	Lack of destination for social gathering	Muscat et al. (2020), Ray et al. (2022), Recknagel and Michener (2008)
4	Population change and socioeconomic change	Kvasničková Stanislavská et al. (2020), Ramirez-Yanez et al. (2018), Rodríguez-García et al. (2019), Rogan and Lacher (2018)
5	Migration and population change	Jackson et al. (2013), Lin et al. (2021), Rusvai et al. (2022), Sawyer (2005), Scholes and Biggs (2005), Seebacher and Post (2015)
Environmental informatics		
1	Effect on agricultural production	Loures et al. (2007), Park (2022), Recknagel and Michener (2008), Semenchuk et al. (2022), Strain et al. (2018), Taylor et al. (2020)
2	Loss of green area and biodiversity	Loures et al. (2007), Threlfall et al. (2017), Tomson et al. (2021), Valerio et al. (2021)
3	Increase of environmental pollution	Paulin (2019b), Recknagel and Michener (2008), Sawyer (2005), Valerio et al. (2021), Wang (2012), Wang (2022)
4	Degrading the local environment	Jackson et al. (2013), Peters et al. (2013), Porter (2018), Valerio et al. (2021)
5	Change in weather pattern and rainfall	Nations (2023), Peters et al. (2013), Recknagel and Michener (2008), Semenchuk et al. (2022)
6	Pollution and traffic congestion	Anthony (2023), Michener (2018), Ray et al. (2022), Recknagel and Michener (2008), Wang (2012)

Results and Discussion

The past studies provide a very strong association between landscape networks, land-use cover and land-use change, and biodiversity, which create different ecological patterns of urban landforms and their networks. This study aims to provide an in-depth picture of these ecological networks by applying different ecological indicators taken as ecological informatics. This network of informatics is related to social, economic, physical, and environmental constraints of an area (Hsu et al., 2022). Among all of these, the environmental constraints are strongly related to the green spaces and biodiversity, which can lead to the fragmentations in landscapes and elimination of green patches. The process of fragmentations is dynamic and changes over time due to natural events and human activities over the years (Ionuţ-Constantin & Eugen, 2016). Therefore, the assessment of this change, along with an understanding of these dynamics morphology of landscape, is needed.

In contrast to this fragmentation, connectivity is necessity for the health of ecosystem and resilience to the environmental change (Martens et al., 2022). This connectivity will shape the ecological network as mentioned above and reduce the isolation. This ecosystem is the combination of social, physical, economic, and ecological components at a given time and scale. Hence, creating this connectivity network is an effective way of reducing the fragmentation and resisting to environmental change; this is considered an effective approach for improving the ecological values of urban areas (Seebacher & Post, 2015).

Conclusions

It is concluded that the ecological values are varying from time to time and area and area (Paulin, 2019b), and the development of these connectivity networks needs a comprehensives approach and research to understand the nature of these networks and issues faced by any particular ecosystem and geographical area (Marselle et al., 2021). However, it has been seen that the urban ecologists faces many challenges and issues during the designing of the morphological connection of these informatics due to limited resources and competing priorities, which may lead to unrealistic results. Biological and ecological sustainability, environmental, and ecosystem play a vital role in ecological sustainability (Wei, 2017). It will ensure and improve the quality of the ecosystem, ecosystem services, and health of the environment, ensure our health, and protect the natural resources for future generations. However, going green and sustainable is challenging in the current era of unplanned development, extensive usage of natural resources, industrialization, and rapid transportation (Wood et al., 2018; Yeh & Liao, 2017). The overuse of oil and gas leads to more emission of carbon and carbon dioxide into the atmosphere, which will lead to an increase in atmospheric temperature and global warming.

To gain an in-depth knowledge of ecological footprint and implementation, it is necessary to study the United Nations Sustainable Development Goals and UN Agenda 2030. This agenda is closely associated with sustainability and ecological development in developed and developing nations for their future. The UN Agenda 2023 is a successor of the Millennium of Goals, which is divided into 17 goals. These goals address the environmental, marine, ecosystem, ecological, social, political, and cross-boundary concerns among the nations for global peace and prosperity. Addressing the needs of the population in health, energy, education, justice, climate, economic growth, human well-being, life on land and under the water, and partnership among the nations, the UN adopted these goals in 2015 (Zulfiqar & Kausar, 2023). The world population is growing rapidly, altering the nature and environment without keeping the ecological balance, particularly the developing states, which creates a gap and shapes the nation as developed and developing. All these scenarios create gaps among the nations, which need to be addressed through a deep examination of the environment, economy, and society with different parameters and cross-cutting research methods and scientific disciplines. This scale and type of research can bridge the gap for global prosperity and justice. In conclusion, for a sustainable future it is necessary to create an eco-friendly society, environmental justice, and a stable economy. This is also a significant need of the day for a better future.

References

Adu-Gyamfi, M., et al. (2021). Effects of internal CSR activities on social performance: The employee perspective. *Sustainability, 13*(11), 6235.

Anthony, B. (2023). The role of community engagement in urban innovation towards the co-creation of smart sustainable cities. *Journal of the Knowledge Economy, 15*, 1592–1624.

Ashrafzadeh, M. R., et al. (2019). Effects of climate change on habitat and connectivity for populations of a vulnerable, endemic salamander in Iran. *Global Ecology and Conservation, 19*, e00637.

Bai, H., et al. (2022). Urban green space planning based on remote sensing and geographic information systems. *Remote Sensing, 14*(17), 4213. https://doi.org/10.3390/rs14174213

Batty, M. (2019). Foreword: The landscape of smart cities. In L. Mora & M. Deakin (Eds.), *Untangling smart cities* (pp. ix–xi). Elsevier.

Bauer, S., & Hoye, B. (2014). Migratory animals couple biodiversity and ecosystem functioning worldwide. *Science, 344*, 1242552.

Bennett, V. J. (2017). Effects of road density and pattern on the conservation of species and biodiversity. *Current Landscape Ecology Reports, 2*(1), 1–11.

Chaudhary, A., et al. (2015). Quantifying land use impacts on biodiversity: Combining species–area models and vulnerability indicators. *Environmental Science & Technology, 49*(16), 9987–9995.

Cook, C. N., & Pinter-Wollman, N. (2019). Social behavior and interactions. In B. Fath (Ed.), *Encyclopedia of ecology* (2nd ed., pp. 319–327). Elsevier.

Crane, M., et al. (2021). Transforming cities for sustainability: A health perspective. *Environment International, 147*, 106366.

Dastjerdi, H. K., & Nasrabadi, N. H. (2022). Interrelationships between urban policy and climate, with emphasis on the environment. *City, Territory and Architecture, 9*(1), 27.

De Montis, A., et al. (2017). Landscape fragmentation in Mediterranean Europe: A comparative approach. *Land Use Policy, 64*, 83–94.

De Palma, A., et al. (2021). Annual changes in the biodiversity intactness index in tropical and subtropical forest biomes, 2001–2012. *Scientific Reports, 11*(1), 20249.

Demaidi, M. N. (2023). *Artificial intelligence national strategy in a developing country*. AI & Society.

Díaz, S., et al. (2019). Biodiversity loss threatens human well-being. *PLoS Biology, 4*(8), e277.

Din, A. U., et al. (2023). How sustainable transportation can utilize climate change technologies to mitigate climate change. *Sustainability, 15*(12), 9710. https://doi.org/10.3390/su15129710

Dudley, N., & Alexander, S. (2017). Agriculture and biodiversity: A review. *Biodiversity, 18*(2–3), 45–49.

Elias, S. A. (2018). Rise of human influence on the world's biota. In D. A. Dellasala & M. I. Goldstein (Eds.), *Encyclopedia of the anthropocene* (pp. 53–65). Elsevier.

García-Vega, D., & Newbold, T. (2020). Assessing the effects of land use on biodiversity in the world's drylands and Mediterranean environments. *Biodiversity and Conservation, 29*(2), 393–408.

Gonthier, D. J., et al. (2014). Biodiversity conservation in agriculture requires a multi-scale approach. *Proceedings of the Royal Society B: Biological Sciences, 281*(1791), 20141358.

González-Marín, P., et al. (2023). Gentrification and its association with health inequalities in Barcelona (2011–2017). *Frontiers in Environmental Health, 2*, 1202149.

Guan, H., et al. (2022). Soil microbial communities in desert grassland around rare earth mine: Diversity, variation, and response patterns. *Sustainability, 14*, 15629. https://doi.org/10.3390/su142315629

Hansen, A., Defries, R., & Turner, W. (2004). *Land use change and biodiversity* (pp. 277–299). Princeton University.

He, C., et al. (2021). A global analysis of the relationship between urbanization and fatalities in earthquake-prone areas. *International Journal of Disaster Risk Science, 12*(6), 805–820.

Hosseini, M., et al. (2019). Landscape connectivity for mammalian megafauna along the Iran-Turkmenistan-Afghanistan borderland. *Journal for Nature Conservation, 52*, 125735.

Houlden, V., Jani, A., & Hong, A. (2021). Is biodiversity of greenspace important for human health and wellbeing? A bibliometric analysis and systematic literature review. *Urban Forestry & Urban Greening, 66*, 127385.

Hsu, Y.-Y., et al. (2022). Beyond the backyard: GIS analysis of public green space accessibility in Australian metropolitan areas. *Sustainability, 14*, 84694. https://doi.org/10.3390/su14084694

Hu, M., & Li, C. (2012). Design smart city based on 3S, internet of things, grid computing and cloud computing technology. In *Internet of things: International workshop*. Springer.

Ionuţ-Constantin, M., & Eugen, P. (2016). Social informatics and the dynamic of contemporary society. In P. Valentina Mihaela (Ed.), *Proceedings of the international conference on interdisciplinary studies (ICIS 2016)*. IntechOpen.

Jackson, L. E., et al. (2013). Agrobiodiversity. In S. A. Levin (Ed.), *Encyclopedia of biodiversity* (2nd ed., pp. 126–135). Academic Press.

Javanmardi, E., Liu, S., & Xie, N. (2023). Exploring the challenges to sustainable development from the perspective of grey systems theory. *Systems, 11*, 20070. https://doi.org/10.3390/systems11020070

Kok, A., et al. (2020). European biodiversity assessments in livestock science: A review of research characteristics and indicators. *Ecological Indicators, 112*, 105902.

Kondratyeva, A., et al. (2020). Urbanization effects on biodiversity revealed by a two-scale analysis of species functional uniqueness vs. redundancy. *Frontiers in Ecology and Evolution, 8*, 73.

Kvasničková Stanislavská, L., et al. (2020). Corporate social responsibility and social media: Comparison between developing and developed countries. *Sustainability, 12*(13), 5255.

Leclère, D., et al. (2020). Bending the curve of terrestrial biodiversity needs an integrated strategy. *Nature, 585*(7826), 551–556.

Lin, L., et al. (2021). Study on equilibrium relationship between government regulation and social environmental responsibility of energy enterprises. *IOP Conference Series: Earth and Environmental Science., 632*, 032028.

Loures, L., Santos, R., & Panagopoulos, T. (2007). Urban parks and sustainable city planning-the case of Portimão, Portugal. *WSEAS Transactions on Environment and Development, 3*, 10.

Malla, F. A., & Bandh, S. A. (2023). Biofuels and sustainable development goals. In K. R. Hakeem et al. (Eds.), *Environmental sustainability of biofuels* (pp. 13–26). Elsevier.

Mamun, M., & An, K.-G. (2022). Key factors determining water quality, fish community dynamics, and the ecological health in an Asian temperate lotic system. *Ecological Informatics, 72*, 101890.

Marselle, M. R., et al. (2021). Biodiversity and health in the urban environment. *Current Environmental Health Reports, 8*(2), 146–156.

Martens, D., et al. (2022). Supporting biodiversity: Structures of participatory actions in urban green spaces. *Frontiers in Sustainable Cities, 4*, 952790.

Meng, X. (2019). Rural–urban migration, urban poverty and inequality, and urbanization in the People's Republic of China. In K. Jayanthakumaran et al. (Eds.), *Internal migration, urbanization and poverty in Asia: Dynamics and interrelationships* (pp. 77–107). Springer.

Meyer, E. A.-O., et al. (2022). The social informatics of knowledge. *Journal of the Association for Information Science and Technology, 70*(4), 307–312.

Michener, W. K. (2018). Project data management planning. In F. Recknagel & W. K. Michener (Eds.), *Ecological informatics: Data management and knowledge discovery* (pp. 13–26). Springer.

Moallemi, E. A., et al. (2020). Achieving the sustainable development goals requires transdisciplinary innovation at the local scale. *One Earth, 3*(3), 300–313.

Molinos, J. G., et al. (2018). Biogeographical shifts and climate change. In D. A. Dellasala & M. I. Goldstein (Eds.), *Encyclopedia of the anthropocene* (pp. 217–228). Elsevier.

Musa, A. A., et al. (2023). Sustainable traffic management for smart cities using internet-of-things-oriented intelligent transportation systems (ITS): Challenges and recommendations. *Sustainability, 15*, 139859. https://doi.org/10.3390/su15139859

Muscat, A., et al. (2020). The battle for biomass: A systematic review of food-feed-fuel competition. *Global Food Security, 25*, 100330.

Nations, U. (2015). *The 17 goals. Sustainable development 2015*. Retrieved from https://sdgs.un.org/goals

Nations, U. (2023). *Popular statistical tables, country (area) and regional profiles UN data set 2023*. Retrieved from http://data.un.org/

Olanrewaju, A. L. A., Yeow, T. S., & Tat, L. L. (2015) *Sustainable housing maintenance management model*.

Park, J. (2022). Effect of external migration on biodiversity in evolutionary dynamics of coupled cyclic competitions. *Chaos, Solitons & Fractals, 158*, 112019.

Pascual, U., Jackson, L. E., & Drucker, A. G. (2013). Economics of agrobiodiversity. In S. A. Levin (Ed.), *Encyclopedia of biodiversity* (2nd ed., pp. 31–44). Academic Press.

Paulin, A. (2019a). Digitalized governance—An embezzled opportunity? In A. Paulin (Ed.), *Smart city governance* (pp. 39–60). Elsevier.

Paulin, A. (2019b). Governing the public apparatus through technology. In A. Paulin (Ed.), *Smart city governance* (pp. 141–142). Elsevier.

Peters, D. P. C., et al. (2013). Landscape diversity. In S. A. Levin (Ed.), *Encyclopedia of biodiversity* (2nd ed., pp. 476–487). Academic Press.

Porter, J. H. (2018). Scientific databases for environmental research. In *Ecological informatics: Data management and knowledge discovery* (pp. 27–53). Springer.

Ramirez-Yanez, L. E., Miller, B., & Reading, R. P. (2018). Conserving biodiversity by restoring ecological processes. In D. A. Dellasala & M. I. Goldstein (Eds.), *Encyclopedia of the anthropocene* (pp. 399–403). Elsevier.

Ray, T., et al. (2022). MODIS-derived fire characteristics and greenhouse gas emissions from cropland residue burning in Central India. *Sustainability, 14*(24), 16612. https://doi.org/10.3390/su142416612

Recknagel, F., & Michener, W. K. (2008). Ecological informatics: An introduction. In *Encyclopedia of ecology*. Elsevier.

Rodríguez-García, A.-M., et al. (2019). Productive, structural and dynamic study of the concept of sustainability in the educational field. *Sustainability, 11*, 205613. https://doi.org/10.3390/su11205613

Rogan, J. E., & Lacher, T. E. (2018). Impacts of habitat loss and fragmentation on terrestrial biodiversity. In *Reference module in earth systems and environmental sciences*. Elsevier.

Rusvai, K., et al. (2022). Changes in the vegetation, soil seed bank and soil properties at bait sites in a protected area of the central European lower montane zone. *Sustainability, 14*, 13134. https://doi.org/10.3390/su142013134

Sawyer, S. (2005). Social informatics: Overview, principles and opportunities. *Bulletin of the American Society for Information Science and Technology, 31*(5), 9–12.

Scholes, R. J., & Biggs, R. (2005). A biodiversity intactness index. *Nature, 434*(7029), 45–49.

Seebacher, F., & Post, E. (2015). Climate change impacts on animal migration. *Climate Change Responses, 2*(1), 5.

Semenchuk, P., et al. (2022). Relative effects of land conversion and land-use intensity on terrestrial vertebrate diversity. *Nature Communications, 13*(1), 615.

Strain, E. M. A., et al. (2018). Eco-engineering urban infrastructure for marine and coastal biodiversity: Which interventions have the greatest ecological benefit? *Journal of Applied Ecology, 55*(1), 426–441.

Taylor, L., et al. (2020). What visitors want from urban parks: Diversity, utility, serendipity. *Frontiers in Environmental Science, 8*, 595620.

Threlfall, C. G., et al. (2017). Increasing biodiversity in urban green spaces through simple vegetation interventions. *Journal of Applied Ecology, 54*(6), 1874–1883.

Tomson, M., et al. (2021). Green infrastructure for air quality improvement in street canyons. *Environment International, 146*, 106288.

Valerio, F., Basile, M., & Balestrieri, R. (2021). The identification of wildlife-vehicle collision hotspots:

Citizen science reveals spatial and temporal patterns. *Ecological Processes, 10*(1), 6.

Wang, M. (2012). Sustainable development. In S. Mannan (Ed.), *Lees' loss prevention in the process industries* (4th ed., pp. 2507–2521). Butterworth-Heinemann.

Wang, Z. (2022). Research on smart city environment design and planning based on internet of things. *Journal of Sensors, 2022*, 2348573.

Wei, F. (2017). Greener urbanization? Changing accessibility to parks in China. *Landscape and Urban Planning, 157*, 542–552.

Wood, E., et al. (2018). Not all green space is created equal: Biodiversity predicts psychological restorative benefits from urban green space. *Frontiers in Psychology, 9*, 2320.

Yeh, J.-C., & Liao, C.-H. (2017). Impact of population and economic growth on carbon emissions in Taiwan using an analytic tool STIRPAT. *Sustainable Environment Research, 27*(1), 41–48.

Zulfiqar, M., & Kausar, M. (2023). Historical development of urban planning theory: Review and comparison of theories in urban planning. *International Journal of Innovations in Science and Technology, 5*, 37–55.

Geographic Information System-Based Solutions for Optimization of Sustainable Solid Waste Management in Gujrat City, Pakistan

Saira Munawar and Imtiaz Ahmad

Abstract

Solid waste is a worldwide complex and growing problem in urban areas. A research study was conducted in Gujrat, Pakistan, by incorporating GIS technique to analyze the efficiency of GIS for sustainable management of solid waste. The study utilizes secondary data acquired from the municipal department that shows the existing infrastructure of solid waste management. The data collection, data cleaning, GIS-based spatial analyses, route optimization, and satellite imagery integration steps are utilized for result development and analyses. The integration of geographic information and spatial analysis empowers stakeholders to make informed decisions, allocate resources efficiently, and implement tailored waste management solutions for the city's diverse needs. It is concluded that it is a dire need of public awareness to use, reuse, and recycle the material as much as possible. This will lead to improving the efficiency of solid waste management. Public awareness in this regard is also necessary for proper management of an area. It has been noted that the educated people can perform the task in a more proper way. Therefore, the general public and local government should corporate and makes parallel efforts at a time for city development and proper solid waste management of an area.

Keywords

Analysis · Data · Development · Gujrat · Human · Solid management · Solid waste · Sustainable solid waste · GIS · Urban

Introduction

It is a bit complex to summarize the word solid waste in a single definition. In wider terms, solid waste is a term according to which any useless or unwanted material that is a by-product of residential, industrial, and commercial activity in any area is a waste (Karunarathne, 2015). Solid waste is of multiple types, which can be categorized according to their origin (domestic, industrial, construction, and institutional) (Kawai & Tasaki, 2016). During the past few decades, a rapid boom in economic development and urbanization has been observed, which significantly improves the living standards of the human being (Tannerfeldt & Ljung, 2012). However, along with this improvement, the consumption of resources has increased rapidly in urban areas due to which the waste released to the environment has increased

S. Munawar · I. Ahmad (✉)
Department of Geography, University of Gujrat, Gujrat, Punjab, Pakistan
e-mail: Saira.munawar@uog.edu.pk

M. M. Anwar, A. Aziz (eds.), *Sustainable Landscape for Urban Resilience*, Sustainable Development Goals Series, https://doi.org/10.1007/978-3-031-99836-2_8

abruptly (Oro et al., 2013). The major reason behind the abrupt increase in the production rate of solid waste is population growth, improving living standards, affluence, and levels of commercial and industrial activities (Panigrahi & Dubey, 2019). Improper handling of solid waste causes air, soil, and water pollution. Nowadays, solid waste has become the most unresolved problem in urban areas and large towns globally, where 80% of the world's population lives and faces a lack of financial resources to deal with the problem (Kumar & Agrawal, 2020). Diseases caused by pollution result in a 10% loss of productive life for each person. In developed countries where most of the generated solid waste is recycled by burning the waste, the ashes produced during burning remain an issue (Nkwachukwu et al., 2013). Pakistan has a population of 231.4 million, with 38.8% people living in urban areas (Muhammad Sarwar Gondal, 2023). Solid waste generated throughout the year is approximately 49.6 million tons (Sialkot Waste Assessment and Characterization Survey, 2010), which is increasing annually at the rate of 2.4%. Gujrat shares 4,499,394 of the total population of Pakistan and contributes 260 tons of solid waste per day. The major challenge faced regarding solid waste management in Pakistan is the lack of reliable data, untrained manpower, shortage of machinery, poor administrative policy, competent private operators, lack of resources, and technical difficulties (Sharma et al., 2020).

Solid waste is generated in tons per day in all areas of the world, which vary in types and quantities (Sadef et al., 2016). These variations in nature and quantity of waste directly depend on the anthropogenic activities and level of technology in a country (Litvinenko, 2020). The more advanced the country, the more complex the solid waste. The problems associated with solid waste are not only because of its large quantity but because of its inappropriate handling and management (Srivastava, Ismail, et al., 2015). The analysis of the waste produced in any area gives a lot of information about its nature, quantity, and trend, which makes it easy for a planner to understand the area's waste generation and handling (Seth et al., 2017). This systematic way of analysis can bring out several factors for the solid waste management application. Geographic information system (GIS) "is a system of computer hardware and software, designed to allow users to collect, manage, analyze, and retrieve large volume of spatially referenced data and associated attribute data collected from a variety of sources" (Jones, 2014). GIS offers multiple geospatial tools that could help administrators to efficiently control manage and deposit solid waste. GIS is one of the modern tools that aid in the decision-making process. It is capable of arranging spatial and nonspatial data in multiple layers and provides tools for waste management planning (Malakahmad et al., 2014).

The study's goal is to examine previous and present solid waste management methods, policies, and practices used by the federal and provincial governments. Another factor to take into account when discussing SWM is population increase, particularly for a country like Pakistan, which has the sixth largest population in the world. For the analysis of the facts of the SWM circumstances that are currently present in the nation, data sources have been used. The amount of trash produced by the residential sector in several cities across the nation has been surveyed as part of a number of studies on SWM. Some nongovernmental organizations (NGOs) and government organizations alike carried out these surveys. In this study, Gujrat, a city in Pakistan's Punjab region that is both industrial and agricultural, has been used as an example city. Both urban and rural residents can be found in the city. Government workers and owners of both small and large private businesses make up the majority of the first group. The majority of the rural population works in the agriculture sector. Gujrat aids in the study of all three forms of waste generation, namely domestic, agricultural, and industrial wastes, as a result of the presence of additional industries in the city. While the data that was examined was of a secondary character, the management of the waste produced by these sectors has been broken down.

Review of Literature

The ever-increasing human population and the anthropogenic activities that go along with it have hastened urbanization in emerging countries over the past 10 years (Avtar et al., 2019). The issue of solid waste has grown significantly, particularly in urban areas. Domestic, industrial, and other wastes—regardless of their low or medium level—have developed into a recurring issue because they keep degrading and polluting the environment (Marshall & Farahbakhsh, 2013). Unsatisfactory waste management practices combined with a hot climate lead to an increase in environmental issues that have substantial local as well as global implications (Marshall & Farahbakhsh, 2013). An effective solid waste management system is urgently needed so that decision-makers and waste management planners can handle the rise in complexity, ambiguity, multiobjectivity, and subjectivity related to this issue (Mir et al., 2021). Despite the growing emphasis on waste minimization at the source, as well as trash recovery and recycling, landfilling remains the most popular technique for disposing of solid waste (Nanda & Berruti, 2021). By spreading the solid waste in thin layers, compacting the solid waste to the smallest practical volume, and covering the solid trash at the end of the operational day, landfills feature an engineering technique of solid waste disposal on land that minimizes environmental risks (Srivastava, Namga, et al., 2015). To ensure its overall sustainability, various important factors, particularly those related to its economics, optimal placement, and operation, must be taken into account, given the rising population density and urban infrastructure. Large tracts of land must be purchased in order to build a municipal solid waste dump, and that area must be located in an urban environment that already has a variety of competing land uses. The preexisting land-use dynamics of the metropolitan area, as well as the type of potential interactions between the landfill and the local environmental, geologic, hydrological, and socioeconomic characteristics, all influence dump location selections. Identification of landfill sites for solid waste disposal remains a crucial management problem in the field of solid waste management, and the choice should be made based on a number of factors (Sumathi et al., 2008). The environment, social progress, and economics can all be significantly impacted by the ineffective execution of waste management policies. The creation of timely waste information is essential to waste policies. In a prior work, we generated in-time waste information using descriptive and predictive analysis (Sabbas et al., 2003). Such real-time data is used to create environmentally friendly policies. Following an assessment of various waste management methods, policymakers take more environmentally desirable concepts for sustainable policymaking into consideration.

Household Waste

The development of sustainable policies takes into account a variety of other aspects, such as the wise use of natural resources and cost-cutting measures, which support the green economy and social progress (Kshitij et al., 2022). The social and economic climate of the area and the viability of the solution with the technology at hand are two obstacles to a sustainable solution in a region (Ahmad et al., 2020). A dynamic balance between economic, social, and environmental factors is stressed as part of the multifaceted notion of sustainable development (Kwatra et al., 2020). Since the 1980s, several other goals and approaches have been incorporated into solid management models. Examples of these objectives include waste generation forecasting, facility site selection, facility capacity growth, facility operating, truck routing, system scheduling, waste flow, and system performance in general. Most of the models mentioned previously had cost minimization as their primary goal. Dynamic models are those that take into account how a system changes over time, while static models are those that do not (Barma & Modibbo, 2022). Three categories of solid waste can be found in a nation like Pakistan, which is classified as a developing nation: residential, agricultural, and industrial trash (Aslam et al., 2022). An estimate states that between one- and two-thirds of the solid garbage produced is not collected. In emerging nations, especially Pakistan, rapid population urbanization has significantly accelerated the problem of solid waste generation. Since improper solid waste management eventually harms the natural ecosystem, solid waste management can be seen as a subset of environmental planning. Every year, a sizeable budget is set aside in each nation for SWM, but the issue persists (Wiltse, 2014). The absence of effective management and technology in the waste collection process is one of the causes. The garbage cannot be properly disposed of after collection, either. Dumping is one of the typical disposal techniques (Mmereki et al., 2017). While dumping grounds are a haven for the breeding of infectious diseases and rats. An entire planned sector H-12 in Islamabad, valued in the billions of rupees, is exploited as a disposal site. This is one of the reasons—among others—that rats and other members of this family can be found in every residence in the city. The priceless land in the capital could be used for housing instead of being dumped, and the money made from this activity might be used to bring about a revolution in SWM conditions across the entire nation (Rahman, 2013). Another factor contributing to migration to the cities is recent flooding in the nation. Most cities lack a formal framework for their structure and can support a particular number of people. Because of this, they are overpopulated, which contributes to the sanitation and SWM issues.

Urbanization and Industrialization

In emerging nations, urbanization has accelerated industrial growth. The urban population receives employment from industries. The trash produced by these industries, on the other hand, has a significant impact on urban SWM as well. The "Plastic Shopping Bag" made of PVC is one of the major industrial solid wastes in emerging nations. One of the concerns, but not the only one, is the cities' worsening solid waste issues. Khyber Pakhtunkhwa is the least urbanized province in the nation, with Sindh being the most urbanized. The trend of urbanization has intensified as a result of the agriculture sector's declining dependence (Arif & Hamid, 2009).

Recycling

After being processed through the proper route, recycling is the act of reusing the waste material (Maloney & Yoxtheimer, 2012). Recycling is not currently practiced consistently throughout the nation. The sector primarily engages in this business on an informal basis and makes money from it. Because there is not any modern equipment or system for recycling in the nation, the economy loses a substantial amount of money every year (Rendon et al., 2021). According to a study on the recycling industry in the city of Lahore, it is a successful industry. Only 21.2% of the city's solid trash is recycled, according to a survey conducted during the study, which generates about Rs. 271 million in revenue. Scavengers (trash collectors) and junk buyers make profits of 15% and 14%, respectively, in this lucrative industry. Recycling has not yet been awarded the status of an industry, but if it is, it may bring in more than 530 million in revenue in the city of Lahore (Batool et al., 2008).

Disposal of Agricultural, Livestock, and Poultry Waste

Due to the country's agrarian economy, trash produced by the agricultural, livestock, and poultry sectors is efficiently consumed or reused throughout the whole nation (Jayathilakan et al., 2012). The primary component of agriculture waste is the raw material produced after crop harvesting. A significant percentage of the trash produced by this industry is utilized as feed for livestock and poultry, such as rice, which is Pakistan's top export and produces 4.4 value added and 0.9 GDP after being harvested (Rehman et al., 2017). Husking is done to remove the seed's outer covering. The husk has a variety of uses, including producing cardboard and feeding livestock. Paper is also made from crop waste that is solidified (Afroze & Sen, 2018).

Dung (excreta) is the primary solid waste produced by the livestock and poultry industries and is disposed of in a variety of methods (Abdel-Shafy & Mansour, 2018). Due to its abundance in organic matter, it is typically employed as manure in the crops (Sharma et al., 2019). Farmers receive subsidies for installation, although the small biogas plant can function on the dung of 5–6 cows or buffaloes. Biogas is a safe gas that has very few negative effects on people's health (Amjid et al., 2011).

Waste Management

Waste management encompasses the processes of "collecting, transporting, processing, recycling, or safely disposing of waste materials." It is important to acknowledge that waste management practices vary significantly between developed and developing nations, urban and rural areas, as well as among residential and industrial waste generators (Rodić & Wilson, 2017). In crafting effective waste management strategies, it is imperative to have a comprehensive understanding of the quantity and characteristics of the generated waste, which forms the foundation for resilient and cost-efficient management approaches (Kumar et al., 2021).

Sustainable Solid Waste Management

The process of waste management comprises many steps, and if these are carried accurately the whole process becomes easier.

1. Waste collection: There are two methods mostly used in the waste collection. In the first type, residents take and dump their waste in municipality dustbins. While in the second method, people put different types of residual material in separate bags and municipality crew pick it up from the doorstep.
2. Waste separation: In general, there are three types of waste mostly generated from households: kitchen waste, nonrecyclable, and recyclable waste. In an efficient waste management system, the residents separate different sorts of waste before dumping or otherwise the workers separate it and treat them accordingly. However, in developing and underdeveloped countries, there is no so such system of treating and recycling different waste types. Eventually, all types of waste are treated equally, which is hazardous considering the infectious material it contains.
3. Waste transportation: Picking up waste from doorsteps, transporting it to the temporary storage sites, and then finally to the dumping site, these steps sum up the transportation phase. Vehicles being used during this process play an important rolc in thc management of a healthy and well-planned system.

In an ideal waste management system, the waste is first transported to temporary storage sites, where useful and recyclable material is separated from trash. In Pakistan, scavengers and other underprivileged people are often involved in selling this waste to small recycling units. This is a way of income for many underaged and overaged workers who are unable to meet their ends by the salaries they get from TMA.

4. Waste treatment: The final step is to treat the waste items according to their properties. The recyclable waste is escorted to the recycling plants. Most of these recycling units in Pakistan are not licensed and not regulated according to the government guidelines and end up making products that are harmful to the general public health.

Biomedical waste is the most crucial one; the Band-Aids, drips, syringes, empty medicine bottles/packs, and surgical instruments that should be treated in a specially designed system are mixed with municipal solid waste. If the plan for perfect solid waste management is followed, then biomedical waste is first incinerated and then dumped (mostly in landfilling) and is never recycled as the chemicals and germs do not become harmless after mere incineration.

All the above mentioned processes are of a good and well-planned waste management system. But this system is perfectly implied in developed countries only. The problem arises when this plan is not accurately executed. In the developing countries, this system is not planned, and the plan is not executed in its exact form.

Geographic Information System Approach

The municipality's efforts fall short of achieving a 100% efficient system and struggle to even reach a 60% level of efficiency. In many areas, municipal bodies are diligently striving to provide their best services for the city's betterment (Schwenkel, 2013). They adhere to their established methods and perform their duties meticulously to run the existing system as smoothly as possible. However, in such situations, it becomes evident that a better and more refined system is necessary, one that takes into account all the relevant facts and figures. This improvement can be accomplished with the assistance of geographic information systems (GIS), which can handle various forms of data, including spatial and attribute data, simultaneously (Burrough et al., 2015).

The existing system appears inadequate in its approach to maintaining a clean environment and is in need of an upgrade and refinement. Solid waste management is a critical area where problems continually arise. As stated by Kumar et al. (2016), "One of the simplest ways to bring innovations to any system is to document and study the existing system and introduce possible reforms through cost-effective solutions." The necessary data set includes information about areas with the highest waste generation, data concerning the employees involved in waste management programs, and a systematic map with comprehensive information about waste generation in different areas, including roads and junctions. The proposed working strategies for municipalities in the Dehradun city situation are as follows.

Implementation of GIS in Solid Waste Management Planning

This approach is based on a case study for Aurangabad, India. It aims to provide a waste management service that is viable within existing financial constraints (Sharma et al., 2020). The action plan comprises two primary elements: the creation of an efficient management information system (MIS) and geographic information system (GIS), and the provision of planning and management strategies to improve financial and institutional support (Hung et al., 2018). Given the financial constraints, the plan suggests incorporating options that can enhance the system without significant capital investment. Therefore, the proposal recommends creating the MIS and GIS information.

Methods

Study Area

Gujrat is one of the oldest centers of civilization in the region. Imperial Gazetteer of India traces its history back to the Mauryan conquest in 332 BC. By eighth century AD, the city was existing as metropolis of considerable importance. The city was given the status of district by the Mughal Emperor Akbar, making it one of the oldest districts of Punjab. In the British era, the city witnessed a lot of war and violence. City's growth spurt started after the birth of Pakistan. The map in Fig. 1 shows the study area.

Data Integration

This section includes the details that were utilized within the study area like the research design, sample collection technique, and data collection modes that are aligned with the research objectives. All steps were written with enough details that ensure the precision and accuracy of this research study. The data collection phase initiated with the collaboration of municipal solid waste management department of Gujrat City. The main aim behind this is to acquire data concerning waste management practices within the city. The data collection process involves collection of information regarding the amount of waste collected each day and diverse types of waste produced from multiple sources within the city. Moreover, details relating to

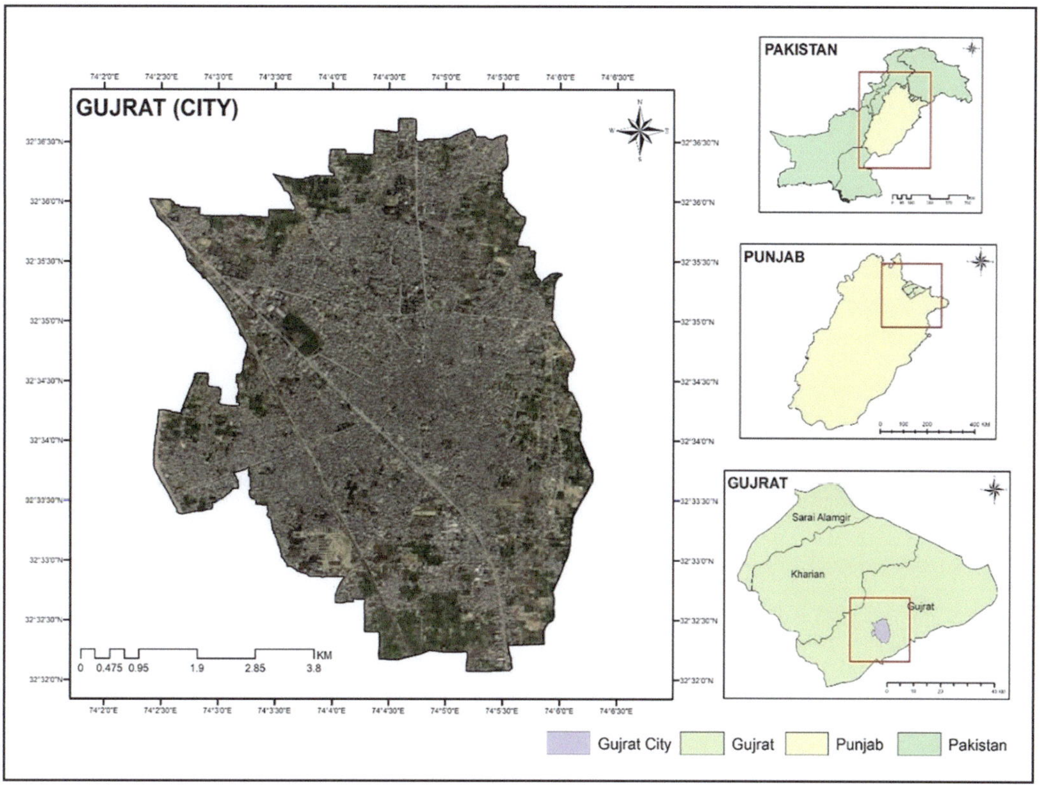

Fig. 1 Map of study, Gujrat, Pakistan

waste management, waste collection schedule, routes of waste-collecting vehicles, and all other methods adopted for waste disposal. Figure 2 provides the framework of methods.

The above acquired data enable a comprehensive understanding of the waste management practice being performed within the city either it is enough to manage the solid waste of the city or not. After the collection of raw data, it passes from the phase where unnecessary information is removed from the data and makes it ready for analyses.

To get a thorough insight into the city's solid waste management system and the problem coming with it, knowing the citizens' perspective was very important. Speaking about the responsible authorities, it was undeniably important to collect information from the workers who are responsible for drainage cleaning and waste collection, the drivers of TMA vehicles and a few high-level officials.

To cater to these points, two questionnaires were developed, one to be filled from general public and the other from TMA workers. District chairman TMA was also informally interviewed, who also shared some valuable information with the researcher. The available data is in the printed format, which also has the information that is not required so before analyses it passes from the data cleaning process. Data cleaning is a process in which unnecessary data is removed. This procedure is important because it enhances the clarity and visualization of the data. After that, the data like collection route, recycling stats, and waste generation data are arranged in Excel sheets for further processing. In this phase, only the required information is taken and the rest of all data is removed.

Before processing data into software like ArcGIS and QGIS, the data is furthered refined and passes from the phase where it arranges and is organized into the formats that are compatible with GIS. This involves the amalgamation of location data with various attributes in a tabular format. This process is facilitated by the "Excel

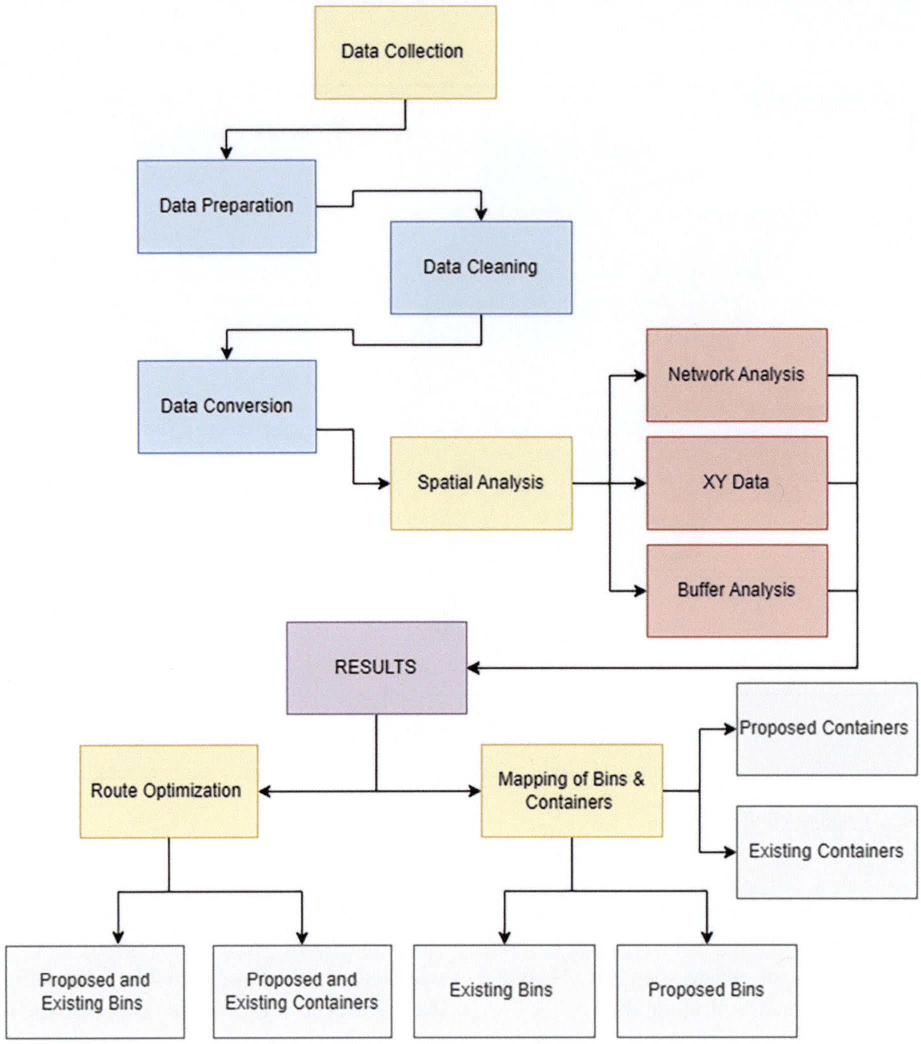

Fig. 2 Flowchart of methodology

to Table" tool, which makes sure the compatibility with the practicality of ArcGIS. Moreover, the spatial context is incorporated into the organized data by displaying XY point data tool. This tool converts the tabular data into feature with coordinates that situate data on the map.

The secondary data of existing waste dust bin is collected from the Municipal Solid Waste Management Department of Gujrat. After calculating the absolute location of the existing dust bins, the data is converted into the point data format that is ultimately overlaid on the map of Gujrat City. This resulting map depicts the distribution and their coverage across various areas of the city. The analyses of the map represent not only the overall existing infrastructure of the city and also highlight the potential gaps and flaws of the plan. It also provides the decision-makers with an opportunity to propose additional plans to address the current infrastructure that falls short, ensuring to enhance more comprehensive waste management facilities in Gujrat City.

The precise mapping of dust locations within the Gujrat City multiple geospatial layers are required, which retain the information like railway lines, road networks, and waterways.

These spatial layers are extracted from the open-street map sources. These layers are helpful in in-depth analyses and broadening the scope of the study.

The extracted layers sustain the important details of the city's transportation setup like the major roads, streets, presence of railways, and waterways. These layers take into account the essential information regarding transportation and accessibility. The strategic planning of waste collection routes and the careful identification of suitable disposal sites are made possible by this all-inclusive approach. The addition of these spatial layers gives the waste management analysis in Gujrat City an additional level of insights through providing a contextual richness.

ArcGIS Pro3 is a GIS-based software that is utilized for the identification of optimize waste collection routes by incorporating the Network Analyst tool. The Network Analyst tool consists of algorithms that calculate the shortest and efficient routes by considering multiple variables like distance between sites, traffic congestion, and other constraints. After the successful implementation of the algorithms, a more efficient waste management system will be developed that reduces fuel consumption, traveling time, labor, and oil cost.

Geospatial analyses and hotspot areas identifications led the researcher to propose new waste collection points within the study area, also taking into account the high waste generation sites and insufficient waste management infrastructure. Positioning of new sites for efficient collection of wastes held with the help of hotspot analyses maps. These maps allow the researcher to allocate waste bins where necessary and aid in optimizing waste management practices, leading to a sustainable waste management system for Gujrat City.

Results and Conclusions

The present study shows a comprehensive result on solid waste management in Gujrat City. The results describe the successful implementation of geospatial analyses. The current study has used advanced GIS technology. The results are combined through GIS, which had given a clear demonstration of Gujrat City's garbage management system. GIS system has given the analysis of existing dust bins' location. Also, it provided the best places to install new dust bins, and optimizing garbage collection and roads information. The abovementioned findings are very helpful for evidence-based decision-making. These findings give a proper way to enhance the sustainability of the garbage management community.

This map in Fig. 3 demonstrates the locations of the city where the dust bins are placed. On the other hand, it also shows a comprehensive visual presentation of current garbage production points. Every point on the map presents the existing bins in the city. This map also represents the clear picture of the roads of the city that are connected for the collection of garbage. The map also gives a spatial view of the arrangements of the bins, which helps the people to discern patterns and concentration. It also highlights the parts of the city that have a high density of garbage collection points. The map has a significant role for the management communities and decision-makers because it gives a real-time and spatial view of network of the bins. Table 1 provides an overview of the distribution of bins, which is helpful for stakeholders to identify the highly populated areas and potential gaps in the garbage collection infrastructure.

There are a number of industries, including the most important Gondal group, Afzal Electronic, and New Book Deport. These are the three top industries of the city. There are 13 other industries mentioned in the list. Latitude and longitude are also shown in the list. This list shows the values of the industries. These findings help in mapping and spatial representation of the industries. For example, the location of Mehar Chowk is 32.57678 north and 74.0896 east. These findings of location help visualize different points in the map, which are significant in the spatial distribution of establishment in the study area. The latitude and longitude of these industries also help in the process of navigation and city planning. It also gives a deep understanding of Gujrat's landscape.

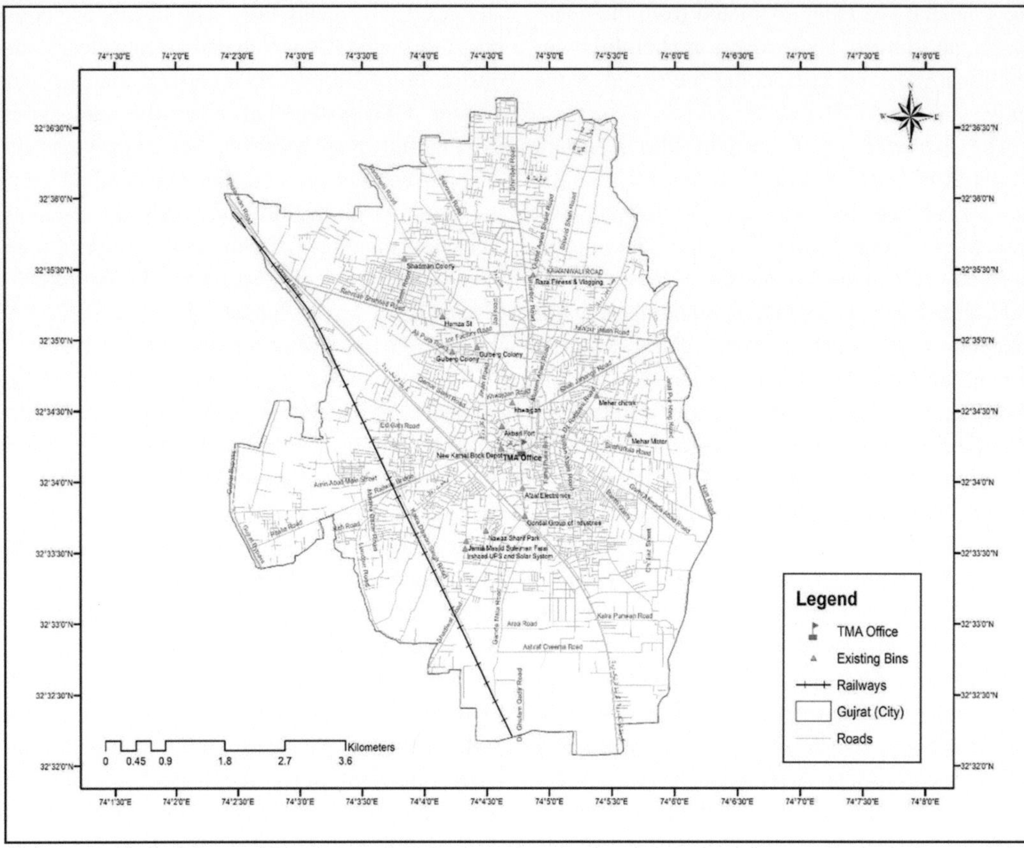

Fig. 3 Location of the existing bins

Table 1 Existing bin locations

S.No.	Name	Latitude	Longitude
1.	Gondal Group of Industries	32.5627	74.08
2.	Afzal Electronics	32.56593	74.07969
3.	New Kamal Book Depot	32.57059	74.07687
4.	Meher Chowk	32.57678	74.0896
5.	Mehar Motors and Rent a Car	32.57223	74.09395
6.	Gulberg Colony	32.58195	74.07034
7.	Gulberg Colony	32.58248	74.07365
8.	Nawaz Sharif Park	32.56092	74.07482
9.	Jamia masjid Suleiman Farsi	32.55974	74.07215
10.	Hamza St	32.58609	74.06902
11.	Raza fitness & Vlogging	32.59101	74.0811
12.	Shadman Colony	32.59291	74.06403
13.	Irshaad UPS and Solar System	32.5588	74.07203
14.	Sunshine Educational Academy	32.55891	74.07442
15.	Southwest Corner of the Akbari Fort	32.57318	74.07698
16.	Khwajgan	32.57595	74.07832

This map in Fig. 4 is a demonstration of the proposed places of dust bins that are identified in the study area, and the place of every proposed dust bins is strategically selected for the geospatial analysis and hotspot identification. This map helps figure out the high trash production areas and poor infrastructure. The present map showcases the proposed dust bins places in various sectors of the study area and also gives a picture of comprehensive overview of waste collection areas. This map provides a spatial pattern of the city, which gives a clear image of waste collection coverage and the areas where bins are mostly needed. The proposed bins places aspire to optimize trash management activities. It also plays a role in contributing in the progress of the city's garbage management system. The current map provides a perfect tool for the management and decision-makers because it helps identify the places that are neglected and need improvement.

These existing garbage containers are purposefully converted to cover different locations, which is including bustling marketplaces, for example, Ajwa Bakery on Bhimber Road, important landmarks such as Nawaz Sharif Park Road, and essential thoroughfares like Sargodha Road Under Bridge. It is the incorporation of specific locations catering to educational institutions, as seen with Islamia School, Zamindar College, and Al Ghazali School. The placement of garbage containers near these schools and colleges emphasizes the commitment to keep a clean and hygienic system for pupil and population alike.

Existing and Proposed Containers

The extensive network of containers expands as shown in Fig. 5; its quantity to cover different huge points, such as the Total Pump on GT

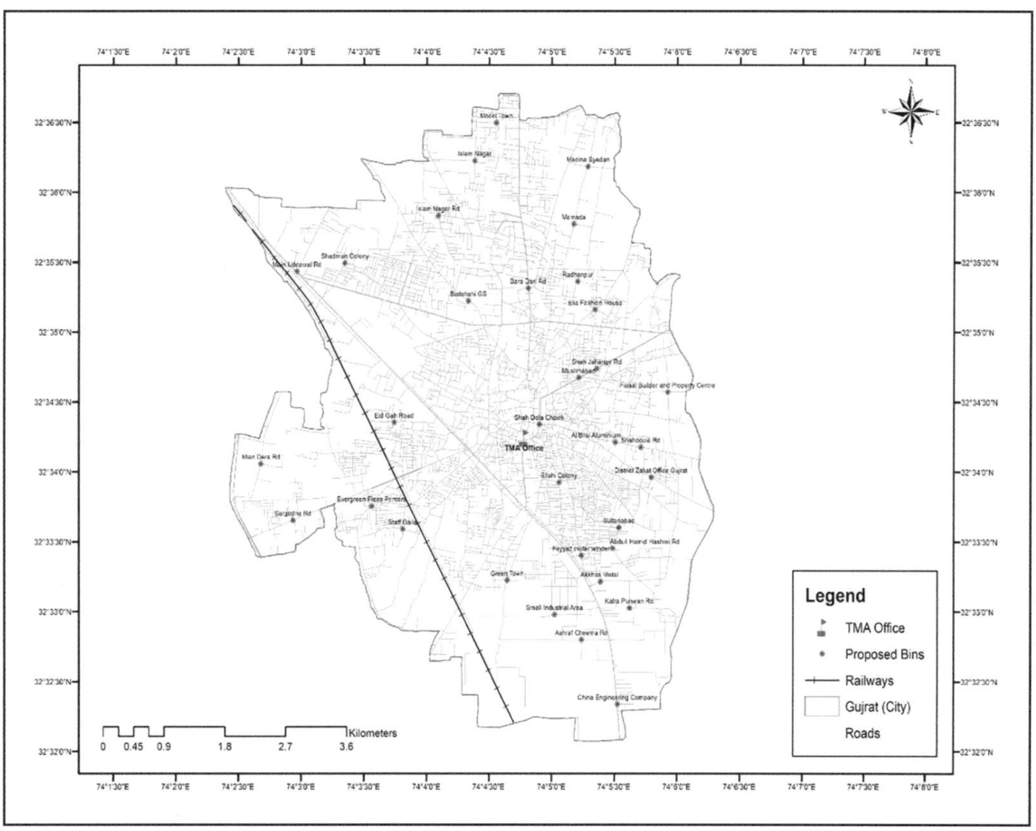

Fig. 4 Proposed bins for urban area

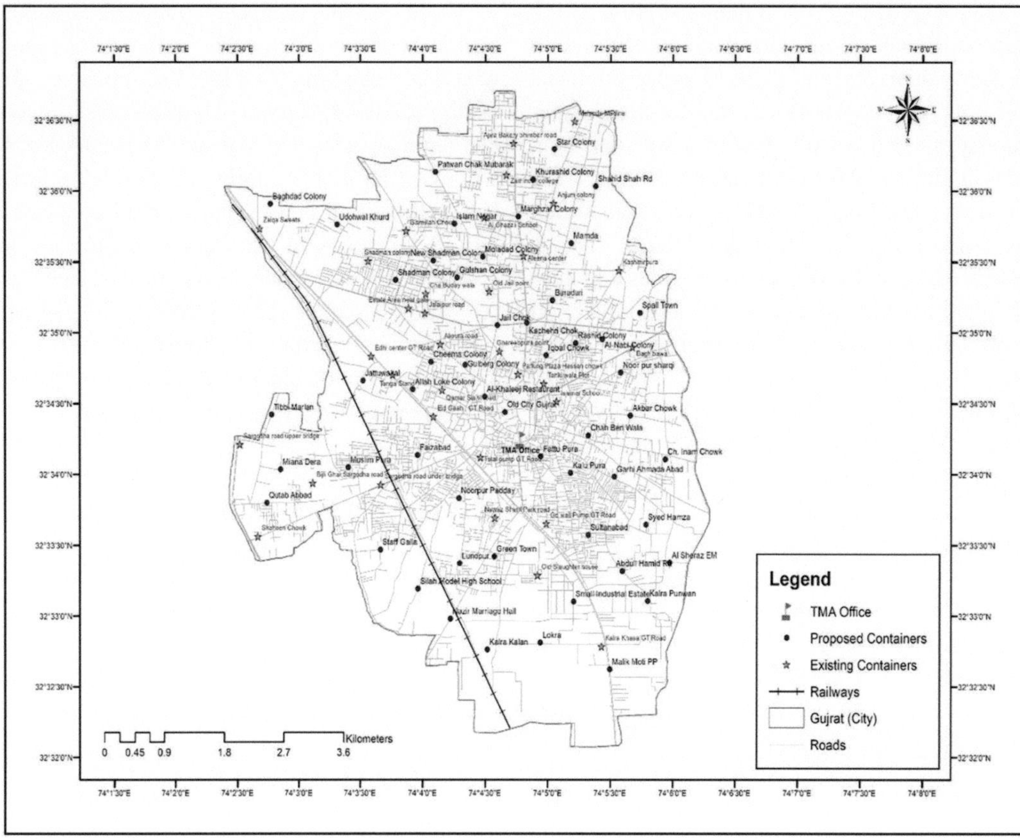

Fig. 5 Proposed and existing containers

Road, Parking Plaza at Hassan Chowk, and the iconic Alipura Road. These containers plays as a basic collection point, which is very much comforting the systematic throw away of waste in link with city planning and garbage management initiatives.

The consolidated map provides a clear view of both existing and proposed solid waste container locations within Gujrat City, presenting important insights into the city's extended garbage management infrastructure. The map visually presents the strategic distribution of present garbage containers; every one of the containers shows specific latitude and longitude coordinates, and also shows the planned places for future containers.

Existing container sites, such as Tanki Wala Plot, Old Slaughter House, and Eid Gaah on GT Road, are juxtaposed with proposed locations like Model Town, Islam Nagar, and Mamada. This juxtaposition permits a comparative analysis of the present garbage collection network and the envisioned expansion, aiding decision-makers in getting coverage and picking up places for improvement in the garbage management system.

Buffer Analysis (Area Coverage—Buffer 200 M)

This map presents the important places of the Tehsil Municipal Corporation office (TMA office), which is the main central location for administrative coordination and overall control of waste management initiatives. This map shows the administrative side of waste management, portraying the city's commitment to organized and effective municipal actions and activities.

This also showcases final overall representation of both existing and proposed container locations, linked with the TMA office point, which provides an extensive spatial understanding for city planners, city officials, and waste management authorities. This extraordinary map helps as an authentic tool for many decision-making, strategic planning, and the future growth of Gujrat City's waste management system to ensure a clean, sustainable, and environmentally conscious urban environment.

The comprehensive map in Fig. 6 provides an inclusive overview of both existing and proposed solid waste container locations within Gujrat City, coordinating spatial analysis techniques to showcase coverage areas and important administrative points. Existing container sites such as Tanki Wala Plot, Old Slaughter House, and Eid Gaah on GT Road are juxtaposed with proposed locations, including Model Town, Islam Nagar, and Mamada.

This map in Fig. 6 employs buffer analysis to showcase the coverage places around every container. In this map, a buffer radius of 200 m has been applied, which creates the parts that encapsulate the approach of every garbage container. This spatial analysis provides a significant understanding of the places served by the existing containers and the extended coverage expected with the addition of proposed locations.

Additionally, the map gives the significant location of the Tehsil Municipal Corporation office (TMA office), serving as a main point for administrative coordination and oversight of garbage management capabilities. This purposeful

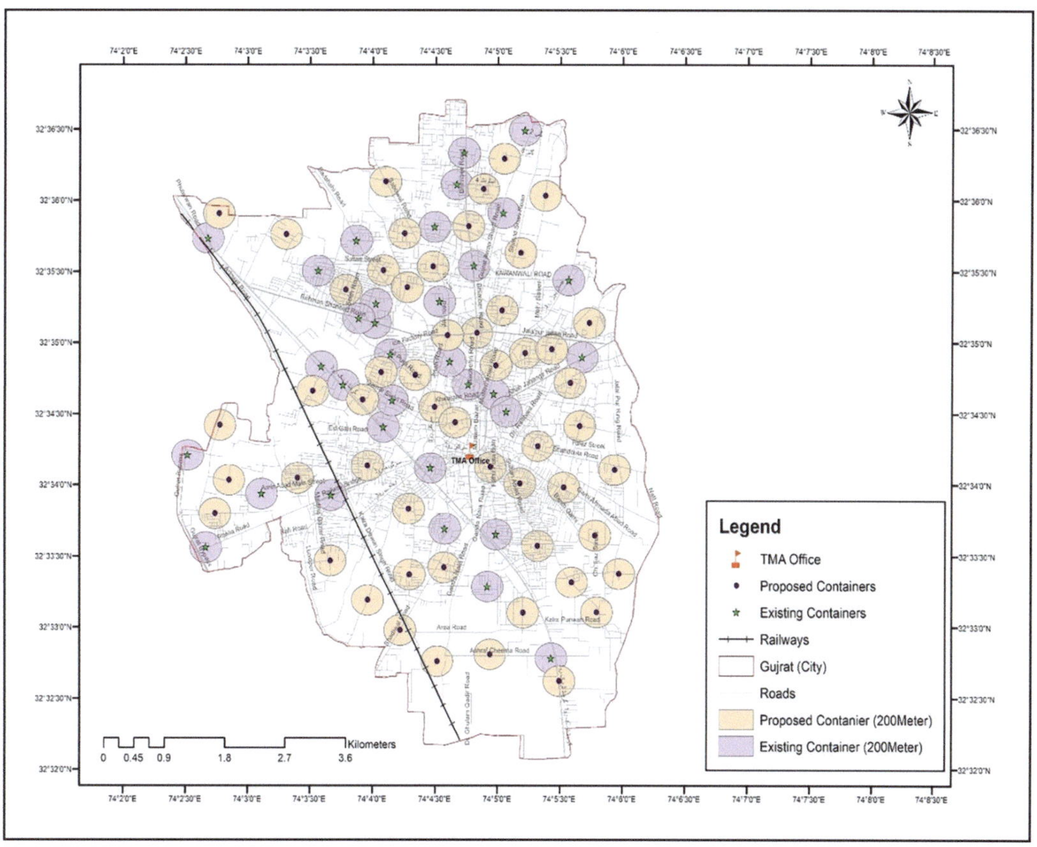

Fig. 6 Buffer analysis (area coverage—buffer 200 m)

involvement highlights the administrative infrastructure that supports the garbage management system because it showcases the city's commitment to assemble governmental and effective urban planning.

This map is a perfect tool because it helps in decision-making. This map incorporates the existing and proposed container places with buffer analysis. The map is beneficial for urban planners, city officials, and waste management authorities who can utilize this extensive representation to get current coverage, identify potential service gaps, and strategically plan for the expansion of the waste management system in Gujrat City. The map also provides valuable insights into the spatial dynamics of waste management; it also supports the efforts to make a cleaner and most perfect, and overall environmentally clean urban environment.

Route from TMA Office to Existing and Proposed Bins

This mentioned map is a true representative the Gujrat City that provides a visual representation of the advanced garbage collection roads from both existing and proposed garbage containers of Gujrat City. The roads in the map are representative of the strategic roads that are taken by garbage collection trucks and containers as they start their travel from the Tehsil Municipal Administration (TMA) office. This point is known as a central point for administrative coordination. After this, these trucks then navigate through different places of the city and collect garbage from assigned containers coming the way.

This map is significant in a way of making the pathways for waste collection vehicles. It also highlights both existing containers, such as Tanki Wala Plot and Old Slaughter House, and proposed containers, including Model Town and Islam Nagar. These advanced roads present a systematic approach to waste collection, which enable the extensive coverage of the city.

The main travel place of these roads is the disposal place, which is known as Boley. The roads emphasize the perfect transportation of assemble garbage to assigned places for proper dispose. These roads begin from the TMA office and then trucks went through the city. They assemble garbage from various containers before going to the final disposal place. The roads and the trucks both contribute to a perfect and efficient trash management activities.

The map in Fig. 7 plays a significant tool for garbage management authorities and the people who are responsible for planning the cleanliness of the city. The present map gives a visual guidance of the advanced garbage collection roads in Gujrat City. It also helps in the system of boosting the effectiveness of garbage management activities that are reducing the environmental bad impact and contributing to the overall cleanliness and in the development of the city.

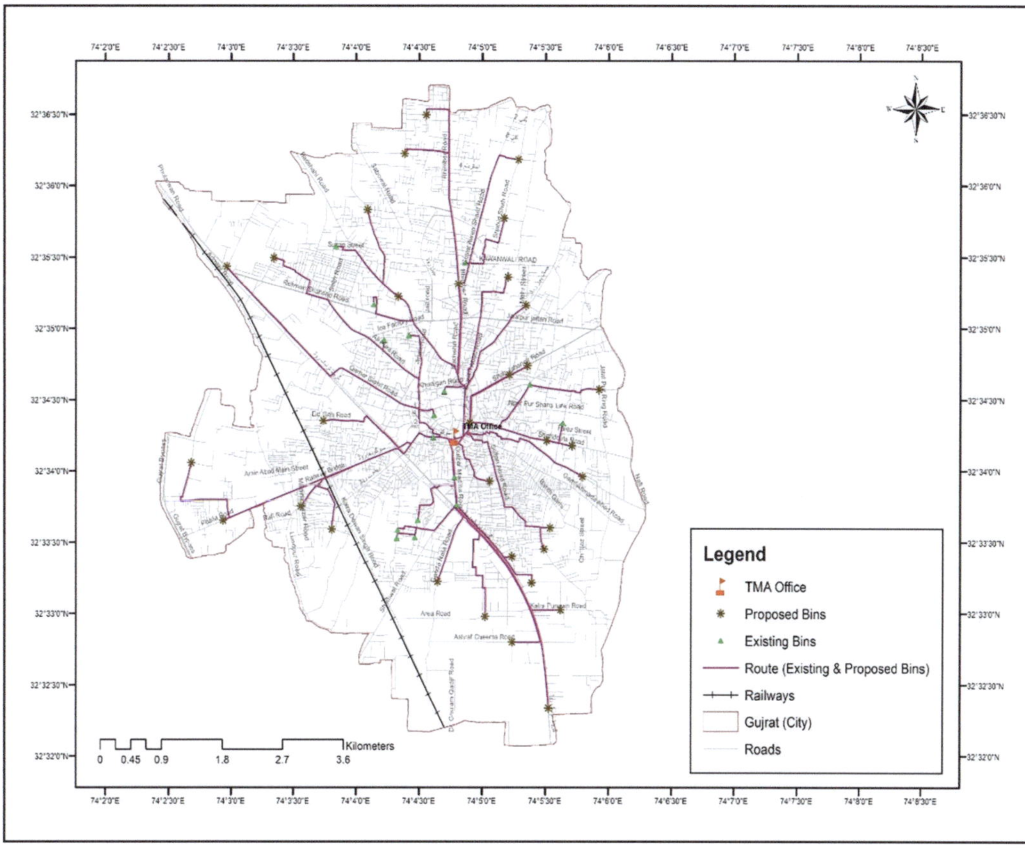

Fig. 7 Route from the TMA office to existing and proposed bins

Route from (Existing and Proposed) Containers to the Disposal Site

This map in Fig. 8 is a true demonstration of a system of optimizing and garbage collection area with perfect roads linked with proposed and existing container location of the study area. This also shows the places of disposal of the waste material, Boley. The map showcases the area of where the containers are placed. There are many huge places like Tanki Wala Plot, Jalalpur Road, Aleena Center, and Nawaz Sharif Park.

The map shows the roads that are extended between the positions of the containers. These roads are interconnected and also work as a network, which is beneficial to cover the main path that goes to the Boley waste material disposal place. The reason of the proposed containers is to build an effective point to cover all waste material. These roads are integrated to the collection of garbage. This also extends the system of cleanliness all over the city.

The most significant place for the disposal of garbage material is known as Boley. It is also known for the final place where all the garbage is collected at the end. It is the central place of the city that is why all other roads are connected with it very effectively. Because it is the main and central place that can help all the mini routes to collect the garbage material and dispose of the material in central position. It also saves time and fuel for the tarpot. This map is designed for the geographic places that helps communicate the system of the garbage disposal approach made for the city.

This map can be used as a great tool for garbage management authorities, city planners, and

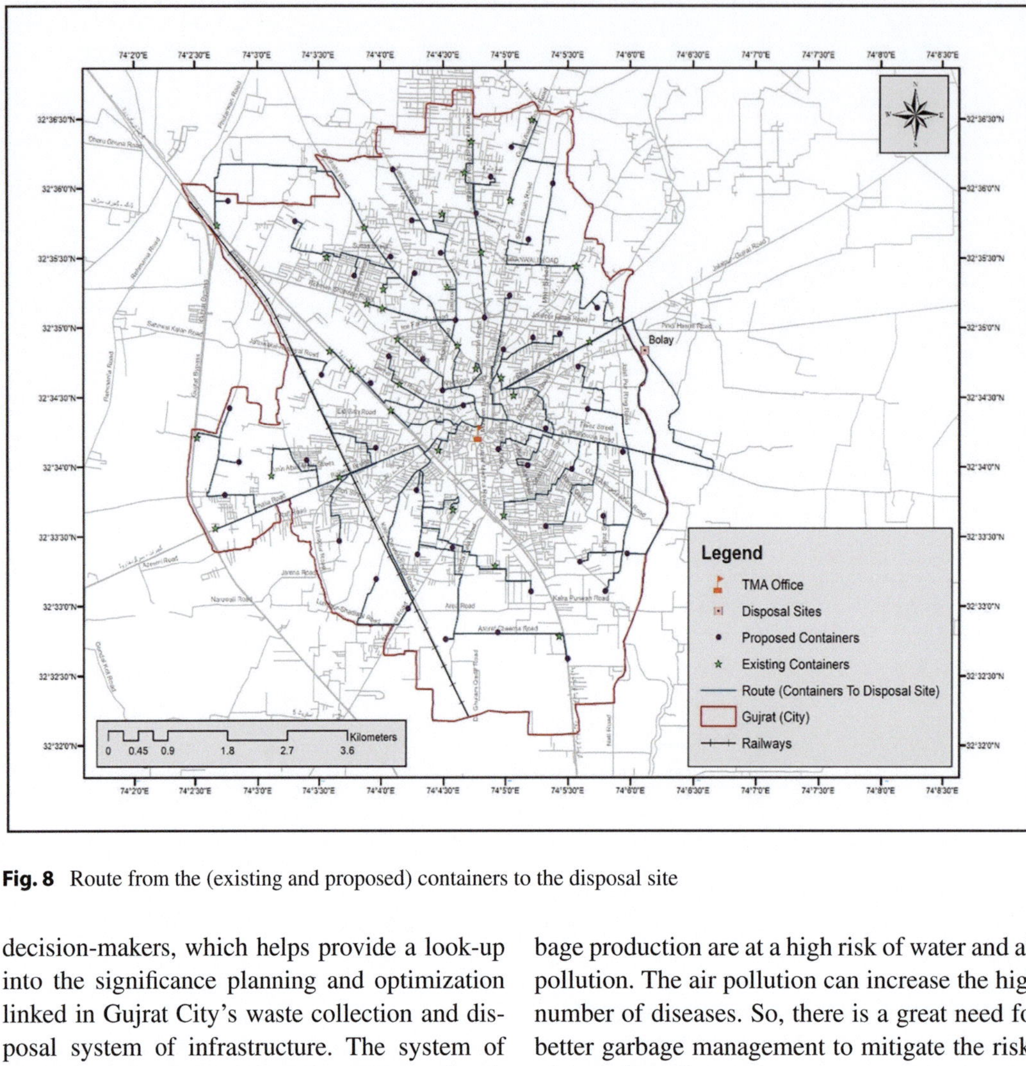

Fig. 8 Route from the (existing and proposed) containers to the disposal site

decision-makers, which helps provide a look-up into the significance planning and optimization linked in Gujrat City's waste collection and disposal system of infrastructure. The system of containers of makes a perspective of future development. This system covers the evolving dynamics of urban waste management and participating to a more peaceful, clean, and disease-free city environment.

Discussion

There is a great need for better roads in high waste production areas. We need garbage processing units and recycling facilities. The city also needs an effective system of garbage collection mechanism. Because if the city is provided with garbage management, we can control the environment and land pollution. The major parts of the city that are having a great number of garbage production are at a high risk of water and air pollution. The air pollution can increase the high number of diseases. So, there is a great need for better garbage management to mitigate the risks of spreading diseases.

Gujrat's waste is not getting deposited properly. The city, due to its rapid industrialization, produces 210.42 tons of solid waste daily. Per capita production stands at 0.5 kg/day, which is likely to rise sharply. A whopping 10% of the solid waste produced in the district is nondegradable plastic, which requires highly developed infrastructure to dispose it off. Unfortunately, the city lacks even a single solid waste treatment plant. The only way of waste disposal is landfill, such as Sheikh Sukha landfill site located 12 km away from Gujrat City.

In summary, the city's solid waste management system requires a comprehensive overhaul, and the integration of GIS can be a pivotal solution in achieving a more efficient and sustainable waste management approach. The maps in the study provide a well-defined for route optimization for city solid waste and its management for city development.

Conclusion

The study has gone through in-depth analyses of solid waste management in Gujrat City. The methodology adopted to assess and produce new plan involves the utilization of advanced GIS-based tools like network analyses that figure out shortest routes based on the factors like traffic congestion and fuel efficiency, also predict new dust bin site by using buffer analyses that highlight the capacity of a single dust bin to facilitate the consumer. By applying the above study, the sustainable waste management of solid can be achieved within the Gujrat City.

The most significant place for the disposal of garbage material is known as Boley. It is also known for the final place where all the garbage is collection at the end. It is the central place of the city that is why all other roads are connected with it very effectively. Because it is the main and central place that can help all the mini routes to collect the garbage material and dispose of the material in central position. It also saves time and fuel for the tarpot. This map is designed for the geographic places that helps communicate the system of the garbage disposal approach made for the city.

References

Abdel-Shafy, H. I., & Mansour, M. S. (2018). Solid waste issue: Sources, composition, disposal, recycling, and valorization. *Egyptian Journal of Petroleum, 27*(4), 1275–1290.

Afroze, S., & Sen, T. K. (2018). A review on heavy metal ions and dye adsorption from water by agricultural solid waste adsorbents. *Water, Air, & Soil Pollution, 229*, 1–50.

Ahmad, S., Iqbal, N., Jamil, F., & Kim, D. (2020). Optimal policy-making for municipal waste management based on predictive model optimization. *IEEE Access, 8*, 218458–218469.

Amjid, S. S., Bilal, M. Q., Nazir, M. S., & Hussain, A. (2011). Biogas, renewable energy resource for Pakistan. *Renewable and Sustainable Energy Reviews, 15*(6), 2833–2837.

Arif, G., & Hamid, S. (2009). Urbanization, city growth and quality of life in Pakistan. *European Journal of Social Sciences, 10*(2), 196–215.

Aslam, B., Maqsoom, A., Tahir, M. D., Ullah, F., Rehman, M. S. U., & Albattah, M. (2022). Identifying and ranking landfill sites for municipal solid waste management: An integrated remote sensing and GIS approach. *Buildings, 12*(5), 605.

Avtar, R., Tripathi, S., Aggarwal, A. K., & Kumar, P. (2019). Population–urbanization–energy nexus: A review. *Resources, 8*(3), 136.

Barma, M., & Modibbo, U. M. (2022). Multiobjective mathematical optimization model for municipal solid waste management with economic analysis of reuse/recycling recovered waste materials. *Journal of Computational and Cognitive Engineering, 1*(3), 122–137.

Batool, S. A., Chaudhry, N., & Majeed, K. (2008). Economic potential of recycling business in Lahore, Pakistan. *Waste Management, 28*(2), 294–298.

Burrough, P. A., McDonnell, R. A., & Lloyd, C. D. (2015). *Principles of geographical information systems*. Oxford University Press.

Hung, N. V., van Hung, P., & Anh, B. T. (2018). Database design for E-governance applications: A framework for the management information systems of the Vietnam Committee for Ethnic Minority Affairs (CEMA). *International Journal of Civil Service Reform and Practice, 3*(1), 98.

Jayathilakan, K., Sultana, K., Radhakrishna, K., & Bawa, A. (2012). Utilization of byproducts and waste materials from meat, poultry and fish processing industries: A review. *Journal of Food Science and Technology, 49*, 278–293.

Jones, C. B. (2014). *Geographical information systems and computer cartography*. Routledge.

Karunarathne, H. (2015). Municipal solid waste management (MSWM) in Sri Lanka. In *Paper presented at the proceedings of the national symposium on real estate management and valuation*.

Kawai, K., & Tasaki, T. (2016). Revisiting estimates of municipal solid waste generation per capita and their reliability. *Journal of Material Cycles and Waste Management, 18*, 1–13.

Kshitij, G., Khanna, N., Yıldırım, Ç. V., Dağlı, S., & Sarıkaya, M. (2022). Resource conservation and sustainable development in the metal cutting industry within the framework of the green economy concept: An overview and case study. *Sustainable Materials and Technologies, 34*, e00507.

Kumar, A., & Agrawal, A. (2020). Recent trends in solid waste management status, challenges, and potential for the future Indian cities–A review. *Current Research in Environmental Sustainability, 2*, 100011.

Kumar, R., Sharma, S., Sharma, S., Kumari, A., Kumar, D., Nadda, G., & Kumar, N. (2016). Chemical composition, cytotoxicity and insecticidal activities of Acorus calamus accessions from the western Himalayas. *Industrial Crops and Products, 94*, 520–527.

Kumar, S., Singh, E., Mishra, R., Kumar, A., & Caucci, S. (2021). Utilization of plastic wastes for sustainable environmental management: A review. *ChemSusChem, 14*(19), 3985–4006.

Kwatra, S., Kumar, A., & Sharma, P. (2020). A critical review of studies related to construction and computation of sustainable development indices. *Ecological Indicators, 112*, 106061.

Litvinenko, V. (2020). Digital economy as a factor in the technological development of the mineral sector. *Natural Resources Research, 29*(3), 1521–1541.

Malakahmad, A., Bakri, P. M., Mokhtar, M. R. M., & Khalil, N. (2014). Solid waste collection routes optimization via GIS techniques in Ipoh city, Malaysia. *Procedia Engineering, 77*, 20–27.

Maloney, K. O., & Yoxtheimer, D. A. (2012). Production and disposal of waste materials from gas and oil extraction from the Marcellus shale play in Pennsylvania. *Environmental Practice, 14*(4), 278–287.

Marshall, R. E., & Farahbakhsh, K. (2013). Systems approaches to integrated solid waste management in developing countries. *Waste Management, 33*(4), 988–1003.

Mir, I. S., Cheema, P. P. S., & Singh, S. P. (2021). Implementation analysis of solid waste management in Ludhiana city of Punjab. *Environmental Challenges, 2*, 100023.

Mmereki, D., Baldwin, A., Li, B., & Liu, M. (2017). Healthcare waste management in Botswana: Storage, collection, treatment and disposal system. *Journal of Material Cycles and Waste Management, 19*, 351–365.

Muhammad Sarwar Gondal. (2023). *Announcement of results of 7th population and housing census-2023*. Retrieved from https://www.pbs.gov.pk/sites/default/files/population/2023/Press%20Release.pdf

Nanda, S., & Berruti, F. (2021). Municipal solid waste management and landfilling technologies: A review. *Environmental Chemistry Letters, 19*, 1433–1456.

Nkwachukwu, O. I., Chima, C. H., Ikenna, A. O., & Albert, L. (2013). Focus on potential environmental issues on plastic world towards a sustainable plastic recycling in developing countries. *International Journal of Industrial Chemistry, 4*, 1–13.

Oro, D., Genovart, M., Tavecchia, G., Fowler, M. S., & Martínez-Abraín, A. (2013). Ecological and evolutionary implications of food subsidies from humans. *Ecology Letters, 16*(12), 1501–1514.

Panigrahi, S., & Dubey, B. K. (2019). A critical review on operating parameters and strategies to improve the biogas yield from anaerobic digestion of organic fraction of municipal solid waste. *Renewable Energy, 143*, 779–797.

Rahman, M. A. (2013). Revisiting solid waste management (SWM): A case study of Pakistan. *International Journal of Scientific Footprints, 1*(1), 33–42.

Rehman, A., Jingdong, L., Chandio, A. A., & Hussain, I. (2017). Livestock production and population census in Pakistan: Determining their relationship with agricultural GDP using econometric analysis. *Information Processing in Agriculture, 4*(2), 168–177.

Rendon, M., Espluga-Trenc, J., & Verd, J.-M. (2021). Assessing the functional relationship between the formal and informal waste systems: A case-study in Catalonia (Spain). *Waste Management, 131*, 483–490.

Rodić, L., & Wilson, D. C. (2017). Resolving governance issues to achieve priority sustainable development goals related to solid waste management in developing countries. *Sustainability, 9*(3), 404.

Sabbas, T., Polettini, A., Pomi, R., Astrup, T., Hjelmar, O., Mostbauer, P., & Speiser, C. (2003). Management of municipal solid waste incineration residues. *Waste Management, 23*(1), 61–88.

Sadef, Y., Nizami, A., Batool, S., Chaudary, M., Ouda, O., Asam, Z.-U.-Z., & Demirbas, A. (2016). Waste-to-energy and recycling value for developing integrated solid waste management plan in Lahore. *Energy Sources, Part B: Economics, Planning, and Policy, 11*(7), 569–579.

Schwenkel, C. (2013). Post/socialist affect: Ruination and reconstruction of the nation in urban Vietnam. *Cultural Anthropology, 28*(2), 252–277.

Seth, D., Seth, N., & Dhariwal, P. (2017). Application of value stream mapping (VSM) for lean and cycle time reduction in complex production environments: A case study. *Production Planning & Control, 28*(5), 398–419.

Sharma, B., Vaish, B., Monika, Singh, U. K., Singh, P., & Singh, R. P. (2019). Recycling of organic wastes in agriculture: An environmental perspective. *International Journal of Environmental Research, 13*, 409–429.

Sharma, H. B., Vanapalli, K. R., Cheela, V. S., Ranjan, V. P., Jaglan, A. K., Dubey, B., & Bhattacharya, J. (2020). Challenges, opportunities, and innovations for effective solid waste management during and post COVID-19 pandemic. *Resources, Conservation and Recycling, 162*, 105052.

Sialkot Waste Assessment and Characterization Survey. (2010). Retrieved from https://www.urbanunit.gov.pk/Download/publications/Files/8/2021/Sialkot-Waste%20Assessment%20&%20Charcterization%20Survey.pdf

Srivastava, P., Namga, S., Singh, P., Arora, N., & Magotra, R. (2015). Hydrothermal chloritisation in granitoids of Chumathang, Ladakh. *International Journal of Earth Sciences and Engineering, 11*, 75–80.

Srivastava, V., Ismail, S. A., Singh, P., & Singh, R. P. (2015). Urban solid waste management in the developing world with emphasis on India: Challenges and opportunities. *Reviews in Environmental Science and Bio/Technology, 14*, 317–337.

Sumathi, V., Natesan, U., & Sarkar, C. (2008). GIS-based approach for optimized siting of municipal solid waste landfill. *Waste Management, 28*(11), 2146–2160.

Tannerfeldt, G., & Ljung, P. (2012). *More urban less poor: An introduction to urban development and management*. Routledge.

Wiltse, J. (2014). The Black–White swimming disparity in America: A deadly legacy of swimming pool discrimination. *Journal of Sport and Social Issues, 38*(4), 366–389.

Harnessing Landsat Data to Quantify the Forest Cover Loss, Land-Use Change, and Temperature Variations: A Case of Bhimber AJK Pakistan

Luqman Subhani, Audil Rashid, Muhammad Mushahid Anwar, and Asad Aziz

Abstract

Background: Human activities are rapidly shrinking the forests worldwide, leading to severe environmental issues. Despite being a country with limited forests, Pakistan has the second highest rate of forest degradation in South Asia. This study investigates the rapid loss of forest cover in Bhimber Azad Kashmir Pakistan, the southwest district of Azad Jammu and Kashmir (AJK), Pakistan, and tracks the associated changes in temperature during the past three decades. The study analysis used mixed-methods approaches, including remote sensing data for geographical analysis and regression statistics for spatial measurement in land-use dynamics. The study used the past 300 years' data to map the land-use land-cover change for forest cover change and its association with land surface temperature. Results: The results of the research confirmed that over the past 30 years, the study area has experienced a gradual loss in forest cover, resulting in a net decrease of 8.22% of the total, which amounts to a total loss of 10,909.84 hectares from 1990 to 2021. Regression coefficient value of 0.83 confirmed a strong association between forest cover loss and land surface temperature, that is, the p value 0.004 less than 0.05 indicates that decrease in forest cover area effectively increases land surface temperature. Conclusions: The study concludes an urgent need to raise awareness among local people about the importance of forest cover and provide valuable insights for policymakers to implement district-level measures in order to safeguard the forests.

Keywords

AJK · Bhimber · Data · Degradation · Human · Pakistan · Prediction · Remote sensing · Forest cover · LST

L. Subhani
Faculty of Science, Department of Botany, University of Gujrat, Gujrat, Pakistan

A. Rashid
Faculty of Science, Department of Botany, University of Gujrat, Gujrat, Pakistan
e-mail: a.rashid@uog.edu.pk

M. M. Anwar · A. Aziz (✉)
Department of Geography, University of Gujrat, Gujrat, Punjab, Pakistan
e-mail: mushahid.anwar@uog.edu.pk; asadaziz@uog.edu.pk

M. M. Anwar, A. Aziz (eds.), *Sustainable Landscape for Urban Resilience*, Sustainable Development Goals Series, https://doi.org/10.1007/978-3-031-99836-2_9

Introduction

Worldwide, forest degradation is amassing at an alarming rate; ultimately forest area is decreasing sharply due to human-damaging activities, including urbanization, land-use changes, and infrastructure development in terms of housing, settlements, road networking, and urban built-up (Maus et al., 2020; Mishra et al., 2022). In spite of being a forest-deficient country, Pakistan experienced a high rate of forest degradation in Azad Kashmir area. National institutes claim that 4.8 percent of total land of Pakistan is covered by forests, but according to the United Nations it is only 2.2%. In South Asia, the rate of deforestation is the highest in Pakistan and demands extensive research. Azad Jammu and Kashmir (AJK) is a state of Pakistan with very diverse types of forests and other plant resources. 42.7% of total area of AJK is covered by forests (Akbar, 2017). These forest types include alpine, subalpine, temperate coniferous, subtropical pine, and scrub forests. Deodar, Kail, Blue Pine, Silver Fir, and Chir-pine plants covered almost 11.6% of total forest area (Shafiq & Mahmood, 2022a). Furthermore, 16.9% of total land area comprised sparsely wooded forests and other trees canopy. Over 90% of the people depend largely upon these forests for their subsistence for food, wood, and fuel. The climate of AJK is adversely affected by forest degradation that ultimately results in a wide range of negative impacts on that area (Akbar, 2017). It has been observed that in the area Azad Kashmir the illegal cutting of trees boosts the forest degradation at economical level for making the profit without any checks and balances or any environmental law. Forest guards have a bit of knowledge about the worth of timber that is illegally traded, and about the other benefits of canopy for land, climate, ecosystem, and environment. The forest cover of Azad Kashmir is lessened by time due to the increasing population and many other agricultural activities that have very adverse impacts on area and its lush green canopy. Due to this alarming rate of deforestation, the climate of AJK is adversely affected, causing a wide range of negative impacts on that area, including flash flooding in the mountains and erode mountainous rocks and larger parts of forest fertile soil (Akbar, 2017).

The study proves that the deforestation directly increases the land surface temperature. In Pakistan, legal wood cutting is four times less than illegal wood cutting, and this illegal activity is increasing day by day to meet the economic needs of the population and urbanization (Maillard et al., 2023). In Pakistan, the areas, including Azad Jammu Kashmir and Khyber Pakhtunkhwa, are most susceptible to degradation of forests by illegal cutting of forests and smuggling of wood by timber mafia. Extensions in farming, cultivations, agriculture, urban area, road networks, and increase in built-up area badly damage the forest cover and water bodies (Khalid et al., 2019).

AJK has a forest cover of 0.567 million hectares that contributes 6.5% to total forest cover of Pakistan according to the Economic Survey of Pakistan 2015–2016. The Pakistan Forest Institute in 2000 conducted a research that demonstrated that forest burnings are fairly frequent in Pakistan; every year, 49,986 hectares of forest out of 3.950 million hectares is destroyed by fire, which makes 1.27% of total forest cover (Mohammad et al., 2024). Another study conducted by Shafiq and Mahmood (2022b) demonstrates that from 2000 to 2020 AJK forest cover decreased by about 46–39%. The drastic change happened from 2005 to 2010 because it is rehabilitation period after earthquake. A study conducted by in Neelum valley AJK shows that prohibited transport of expensive wood and irresponsibility, carelessness, corruption, and ineffectiveness of AJK forest department is the main cause of forest degradation rather than people using wood for fuel purposes; this study also demonstrated that indigenous people do not cut trees for their livelihood assistance.

Forests play a vital role in controlling land surface temperature because they have lower albedo surface and higher evapo-transpiration (Ellison et al., 2017; Prevedello et al., 2019). Another study highlights that forest degradation and removal of green vegetation result in an increase in land surface temperature. Radiant surface temperature greatly exceeds as soil surface is exposed due to the removal of forest covering. Landsat images are a very effective means of mapping forest cover and forest changes in Pakistan at temporal and spatial scale, which use remote sensing data. For accurate mapping of forest cover, use of high-resolution Landsat data is very rare in Pakistan.

In this study, "kappa statistic" was performed to validatc thc classification of image. Dot matrix is frequently used for the determination of pre-pixel image classification. It is very compulsory to use high-resolution data for effective and accurate measurement of deforestation in Pakistan (Ahmad et al., 2021). Furthermore, it will be very helpful for the Ministry of Climate Change to actively implement on UN REDD+ initiative. Excel forecasting method is very easy to operate, fast, reliable, and does not require long-term data to forecast (Rahardja, 2021).

In Pakistan, only one national-level inventory for forest tracking is done so for by conventional method; instead, use of remote sensing technique in 1992 and after that only 73 peer-reviewed research articles were published on forest mapping. Out of 73, only 3 studies were conducted so for on AJK forests (Ahmad et al., 2021). Ignorance of forest department, shortage of detailed and comprehensive research work to highlight the actual status of AJK forests, demands an advanced reliable forest cover tracking and associated temperature variations. Therefore, the objective of this study is to determine the rate of deforestation in Bhimber AJK and associated temperature variations over the past three decades and predict the forest's status and temperature scenario up to 2050 on the basis of past trends by using remote sensing data.

Research Methodology

Study Area

District Bhimber lies between latitude 32.58 N and longitude 74.04 E with elevation of 1005 feet above sea level. District Bhimber is bounded by district Kotli in the north, district Mirpur in the northwest, Gujrat Punjab in the west and south, and India-Occupied Kashmir in the east. District Bhimber is located at the extreme south of the state of AJ&K; hence, the climate of the district is generally hot and dry. Figure 1 shows the map of study area with elevation from sea level. The average annual temperature is 33.5 °C, with an annual average rainfall of 281 mm. District Bhimber is blessed with dense lush green forests.

Out of the 1.007 million hectares of conifers of the state of AJ&K, 0.059 million hectares are standing in this district, along with 0.004 million hectares of riverine bela. A detailed methodology of research work elaborated in the form of a flow chart is shown in Fig. 2. The process involved the conceptualization of the topic, research design, data collection, and methods. The methods involved the techniques of geographical information system and remote sensing.

Data Acquisition and Preprocessing

For forest mapping and temperature measurements of study area district wise, Landsat 5 TM, 7 ETM+, and Landsat 8OLI satellite images of years (1990, 1995, 2001, 2005, 2010, 2015, and 2021) with resolution of 30 × 30 m were downloaded from the "United States Geological Survey" official website (https://www.earthexplorer.usgs.com). Details of data acquisition and source are represented in Table 1. Image preprocessing includes radiometric correction, atmospheric correction, and geometric correction done after downloading images to obtain high-

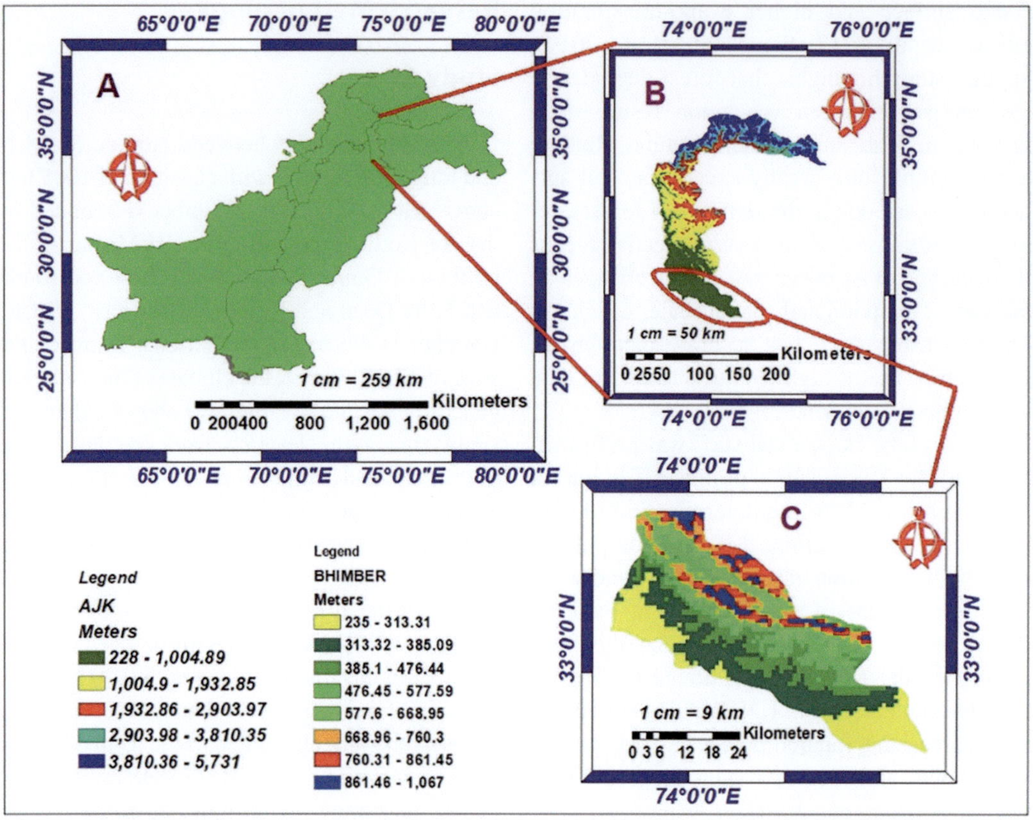

Fig. 1 Map of study area with elevation from sea level. (**a**) Map of Pakistan, (**b**) map of Azad Jammu and Kashmir (AJK), and (**c**) map of study area district Bhimber

quality and clear images. In May, only permanent trees and coniferous forest remain; there is no crop planted in this time period, so this time period is selected.

Data Collection

For processing of images for land use and land classification to measure forest cover changes, these were transferred to Arc GIS 10.7.1. Landsat 5TM, and 7ETM that contain 7 bands and Landsat 8OLI images comprised 11 bands, and band composition was performed using the Arc toolbar to create a single composite raster file. After this, layer stacking was done by assigning 543 for Landsat 5TM, 7ETM raster files, and 432 for Landsat 8OLI to red, green, and blue band, respectively, for the creation of false band combination to distinguish between vegetation, water, and built up with greater accuracy. The area of interest (respective district) is made by "Arc extract by mask" toolbar by masking district boundary-shaped file obtained from https://www.diva.gis.com/. For land-use and land-cover mapping, interactive supervised image classification has been performed to distinguish between forest cover, water bodies, barren soil, built-up area, and snow cover on the basis of ground truth data. Forty-five samples for each attribute were taken for interactive supervised classification and then merged to their respective class. After this, from the classification toolbar interactive supervised classification was operated to obtain classified image.

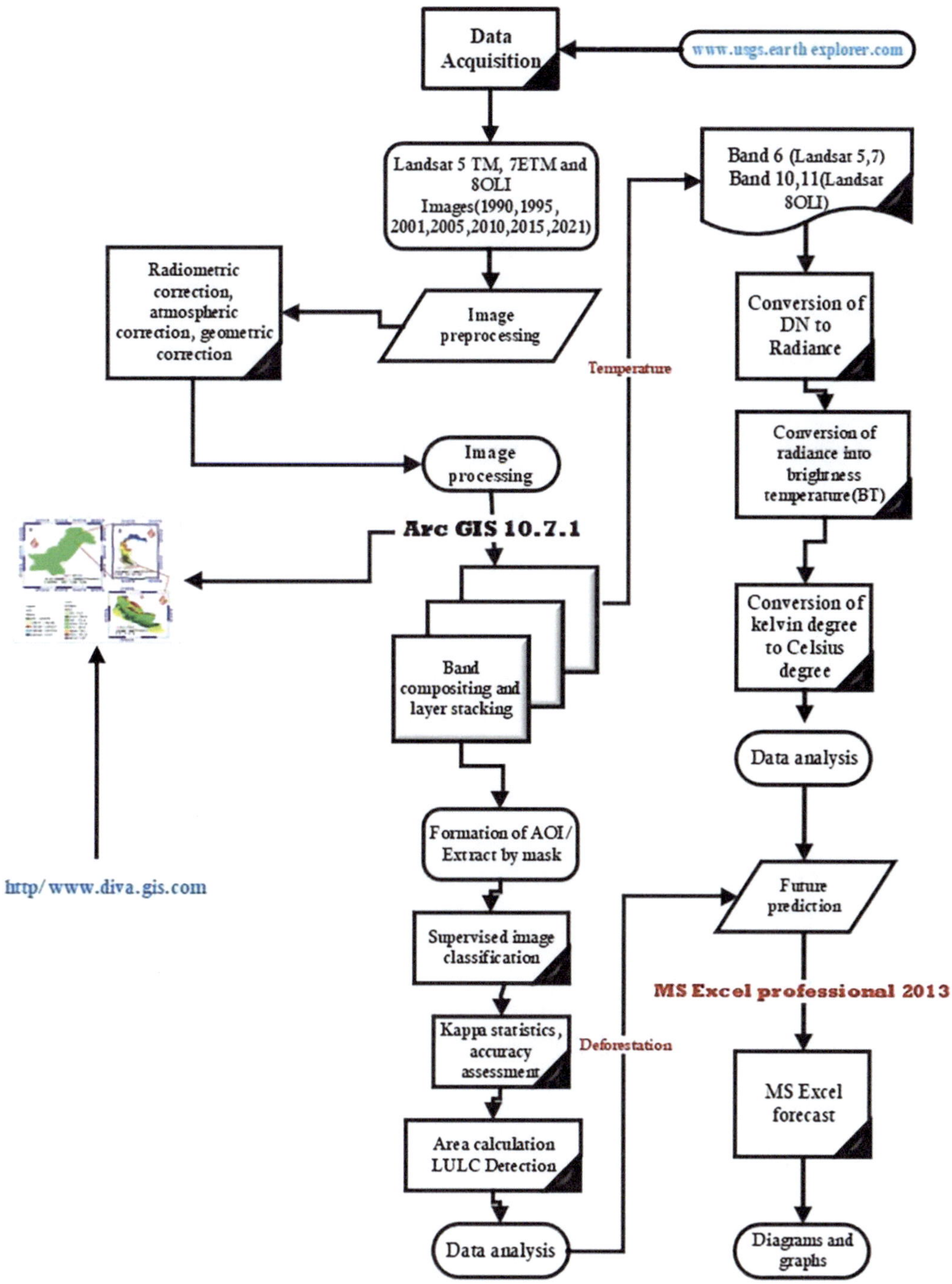

Fig. 2 Research methodology flow chart

Table 1 Image acquisition and arrangement

Year	Landsat	Resolution (m)	Date	Source
1990	5 TM	30 × 30	4/28/1990	https://www.earthexplorer.usgs.com
1995	5 TM	30 × 30	4/21/1995	https://www.earthexplorer.usgs.com
2001	5 TM	30 × 30	4/29/2001	https://www.earthexplorer.usgs.com
2005	5 TM	30 × 30	4/23/2005	https://www.earthexplorer.usgs.com
2010	5 TM	30 × 30	4/18/2010	https://www.earthexplorer.usgs.com
2015	7 ETM+	30 × 30	5/2/2015	https://www.earthexplorer.usgs.com
2021	8 OLI	30 × 30	5/3/2021	https://www.earthexplorer.usgs.com

Area Calculation (Forest Cover)

Before the calculation of area for each class, raster file was converted to vector polygon format. After this, a new field was added to the polygon file attribute table, namely "area". The calculate geometry tool area was calculated for each attribute in hectares. The attribute table was then exported to MS Excel in the form of dbase file, and the area for each class was calculated by inserting class name and area values in the pivot table. The area for each class designated was measured in hectares.

Kappa Statistics and Accuracy Assessment

The last step of image classification is to check the validity of data and accuracy of classification. Kappa statistic and accuracy assessment were performed to check the validity and accuracy of image classification. Error or confusion dot matrix method were used for this purpose; randomly 120 points were selected on classified image and then converted to shape file and saved after this image is cross-referenced with Google Earth data. The data was then organized in table in row and column form to check the producer and user accuracy. The confusion matrix was constructed for all images on the basis of best speculation of their illustrative class. The kappa statistic was performed to validate the classification of image. Dot matrix is frequently used for the determination of pre-pixel image classification. Kappa coefficient values higher than 0.80 and up to 1 indicate that image classification is accurate and reliable. Hussain and Karuppannan (2023) calculated the accuracy assessment of supervised classified images by using the following formulas.

$$\text{The overall classification accuracy} = \text{No.of correct classified points} / \text{total number of reference points} \times 100$$

Abubakar et al., Hussain and Karuppannan (2023), used the following formula for the calculation of kappa coefficient:

$$\text{Kappa}(\text{K}) = \text{Observed Accuracy} - \text{chance agreement} / 1 - \text{Chance agreement}$$

Land Surface Temperature Measurements

For temperature measurement, same data sets were used as those listed above. Band 6 (thermal band) from Landsat 5 TM, Landsat 7 ETM, and bands 10, 11 (thermal band) from Landsat 8OLI of all selected images were used and processed in Arc GIS separately. After the extraction of area of interest (respective district) from the Arc GIS extract by mask tool bar band that was processed further using the spatial analyst toolbar. Using raster calculator, land surface temperature (LST) was measured by using the following method. For the calculation of land surface temperature from Landsat 5 and 7 (band 6), the following method was used:

Step #1: Conversion of the DN to spectral radiance (L) by using Eq. (1)

$$L(\lambda) = \left(\frac{LMAX - LMIN}{QCALMAX - QCALMIN} \right) \times (QCAL - QCALMIN) + LMIN \quad (1)$$

LMAX maximum radiance (Wm^{-2} sr^{-1} μm^{-1}), *LMIN* minimum radiance (Wm^{-2} sr^{-1} μm^{-1}), *QCAL* DN value of pixel, *QCALMAX* maximum DN value of pixels, *QCALMIN* minimum DN value of pixels.

For Landsat 8OLI (band 10, 11), spectral radiance was calculated by Richard and Abah (2019) using Eq. (2):

$$L\lambda = ML \times Qcal + AL \quad (2)$$

where Lλ is the spectral radiance in (Wm^{-2} Sr^{-1}), *ML* the radiance multiplicative scaling factor for the band, *QCAL* DN value of pixel, *AL* radiance additive scaling factor for the band.

Step# 2: Conversion of DN to brightness temperature (BT) by using Eq. (3)

$$T = K2 / \ln(K1 / L(\lambda)) + 1 \quad (3)$$

where K1 and K2 are calibration constants.

Step#3 is the conversion of Kelvin temp to degree Celsius by using Eq. (4)

$$Temp(centigrade) = T - 273.35 \quad (4)$$

Effect of change in forest area and settlements on land surface temperature interpreted by using linear multiple regression in MS Excel by taking land surface temperature as dependent variable Y and vegetation cover- and built-up area as independent variables as X.

Future Predictions

Forest mapping data obtained from LULC and land surface temperature (LST) data of the past three decades were further analyzed for future prediction up to 2050 by MS Excel forecast method in MS Excel Professional 2013. The obtained results are then presented in the form of tables and graphs for the evaluation of research objectives. Rahardja (2021) used the following Excel formula for forecasting of data.

$$= FORECAST(x, known\_y's, known\_x's)$$

X Required the data point for which you want to predict a value, *Known_y's required* the dependent array or range of data, *Known_x's required* the independent array or range of data.

Results

Supervised Image Classification

For LULC of Bhimber district, land surface is classified as vegetation cover, barren soil, built-up area, rock surface, and water bodies. The results of LULC for Bhimber district are represented in Table 2. Land use and land classification for mapping vegetation cover changes of Bhimber district demonstrated a drastic decrease in vegetation cover from 1990 to 2021. As results indicate, from 1990 to 2021, vegetation cover decreased from 48,219.93 (ha), that is, 35.82% to 37,310.09 (ha), that is, 27.60%, a net change of −8.64% in vegetation cover noticed in respective years.

Furthermore, built-up area increased from 2.34% to 12.28% with net change of 9.94% in increasing order. A sudden increase in rock surface area was observed from 2010 to 2021, an increase of 9.86% recorded in just 11 years due to the burning of vegetation in fire season (Fig. 3). The results of image accuracy and kappa statistics are presented in Table 3, showing that the values of kappa coefficient for all year's images were greater than 0.80 and nearest to 1, proving the validity and accuracy of image classification.

Table 2 LULC table of Bhimber AJK district from 1990 to 2021

Feature	Area (ha)	%	Area (ha)	%	Area (ha)	%	Area (ha)	%	Area (ha)	%	Area (ha)	%	Area (ha)	%
	1990		1995		2001		2005		2010		2015		2021	
Barren soil	63,539.74	47.01	56,579.41	41.86	43,344.8	32.06	42,557.8	31.48	32,044.60	23.70	17,852.24	13.20	26,952.24	19.48
Built-up area	3169.59	2.34	3268.68	2.41	9019.15	6.67	9238.08	6.83	11,273.09	8.34	12,300.51	9.10	16,600.5	12.28
Rock surface	16,397.92	12.50	30,663.26	22.68	38,468.86	28.46	38,907.65	28.78	48,947.98	36.21	63,057.83	46.65	51,257.83	38.32
Forest cover	48,219.93	35.82	42,714.6	31.30	42,539.32	30.73	42,330.39	31.24	38,463.95	28.64	38,520.09	28.60	37,310.09	27.60
Water bodies	3131.54	2.31	3332.1	2.46	2887.67	2.13	2225.19	1.64	4230.54	3.13	3280.43	2.427	3140.53	2.32

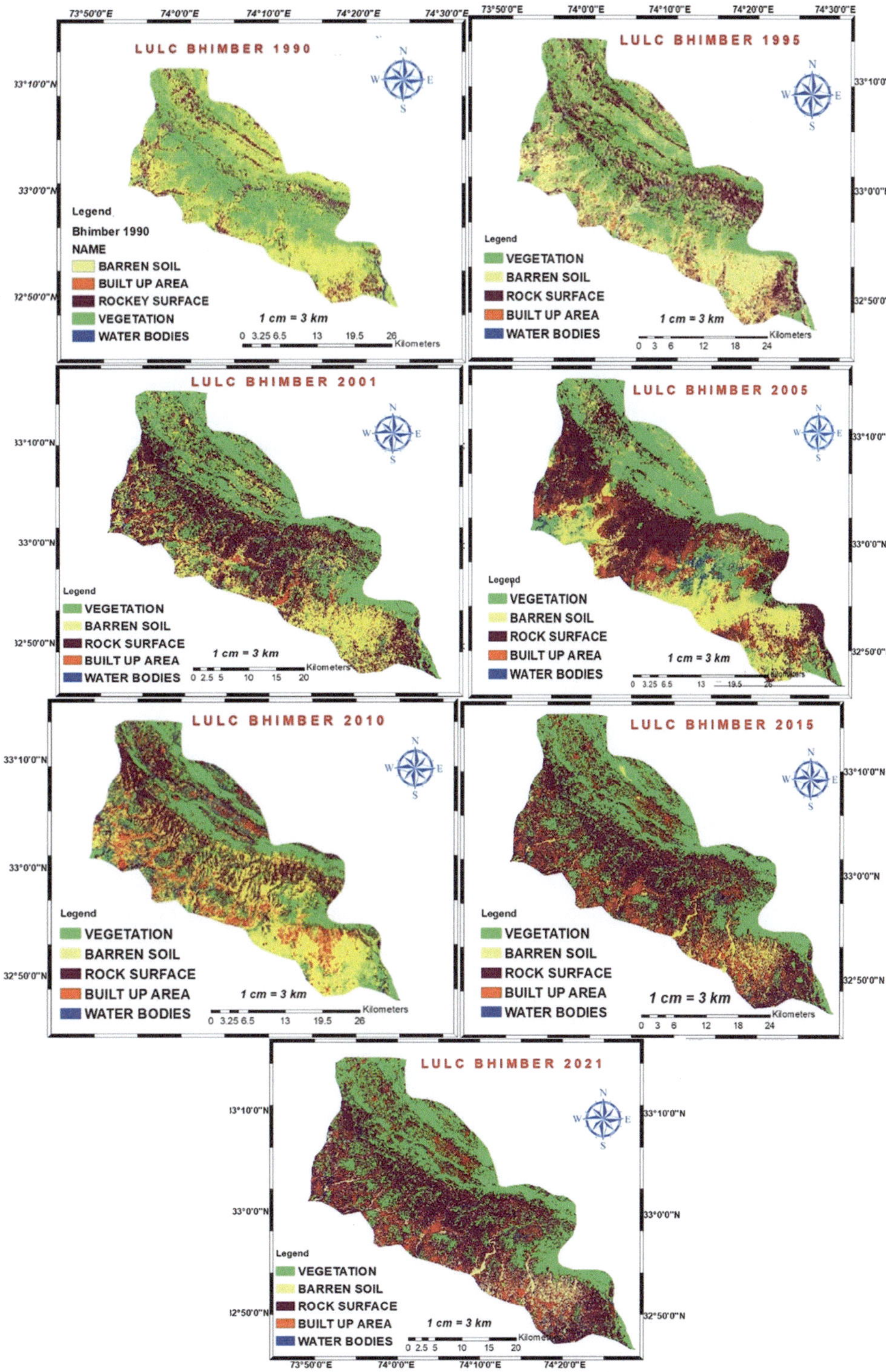

Fig. 3 LULC maps of District Bhimber from 1990 to 2021

Table 3 Overall classification accuracy and kappa coefficient

	1990	1995	2001	2005	2010	2015	2021
Accuracy assessment (%)	94.65	93.60	89.90	91.80	95.75	95.50	94.85
Kappa coefficient	0.934	0.916	0.876	0.908	0.923	0.942	0.931

Table 4 LST values for District Bhimber from 1990 to 2021

	1990	1995	2001	2005	2010	2015	2021
Minimum value (°C)	17.41	17.81	17.66	20.79	7.84	9.54	8.54
Maximum value (°C)	41.6	42.5	43.08	43.94	43.97	44.10	44.5

Land Surface Temperature Measurements

Land surface temperature from 1990 to 2001 presented in Table 4 results shows that there is a continuous increase in temperature upper value from 41.6 °C to 44.5 °C of years 1990–2021. An increase of 2.9 °C in upper temperature value observed in the respective time period demonstrated a rise of 0.99 °C per decade (Fig. 4). The random trend in lower temperature values was observed. Table values show an increase of lowest value from 1990 to 2005. The lowest temperature value was observed in 2010, that is, 7.84 °C. The trend in temperature scenario is presented in Fig. 5. Land surface temperature (LST) maps of Fig. 4 show the land surface temperature variations in the respective study area.

Association Between Vegetation and Land Surface Temperature

The effects of vegetation changes on temperature scenario are predicted by linear regression model by taking temperature as a dependent factor and vegetation cover as an independent factor.

Regression analysis results showed that the R square value is 0.83, which is very close to 1, indicating that vegetation changes mean decrease in forest cover effectively increasing the land surface temperature. The R square value showed that 83% of the dependent variable is explained by independent variable. From the ANOVA table, the value of "P" is 0.004, that is, less than 0.05, so regression results clearly showed that a decrease in vegetation significantly resulted in an increase in land surface temperature. Figure 6 shows the graphical representation of regression analysis with continuous increase in land surface temperature with decrease in vegetation.

Future Prediction Up to 2050 for Forest Cover and Temperature

The Excel forecast data for vegetation and temperature scenarios up to 2050 is presented in Table 5 on the basis of the past three decades' data. Results demonstrated that the forest cover of respected area will decrease to 23.5% up to 2050 from 28.6%, which was in 2021. From 1990 to 2021, the decrease in forest cover area was 8.64%; and from 2021 to 2050, the decrease in forest cover area was 5.75% as forecasted. According to the table values, for temperature forecast an increase of 2.36 °C was forecasted from 2021 to 2050. Figures 7 and 8 show the future prediction trends for District Bhimber up to 2050 for forest cover and land surface temperature (LST), respectively.

Discussion

The results of the study confirmed a continuous decrease in forest cover by 8.64%, which amounts to a total loss of 10,909.84 hectares from 1990 to 2021 in the respective study area. Over the past three decades, this ultimately resulted in a rise of land surface temperature by 2.9 °C due to the

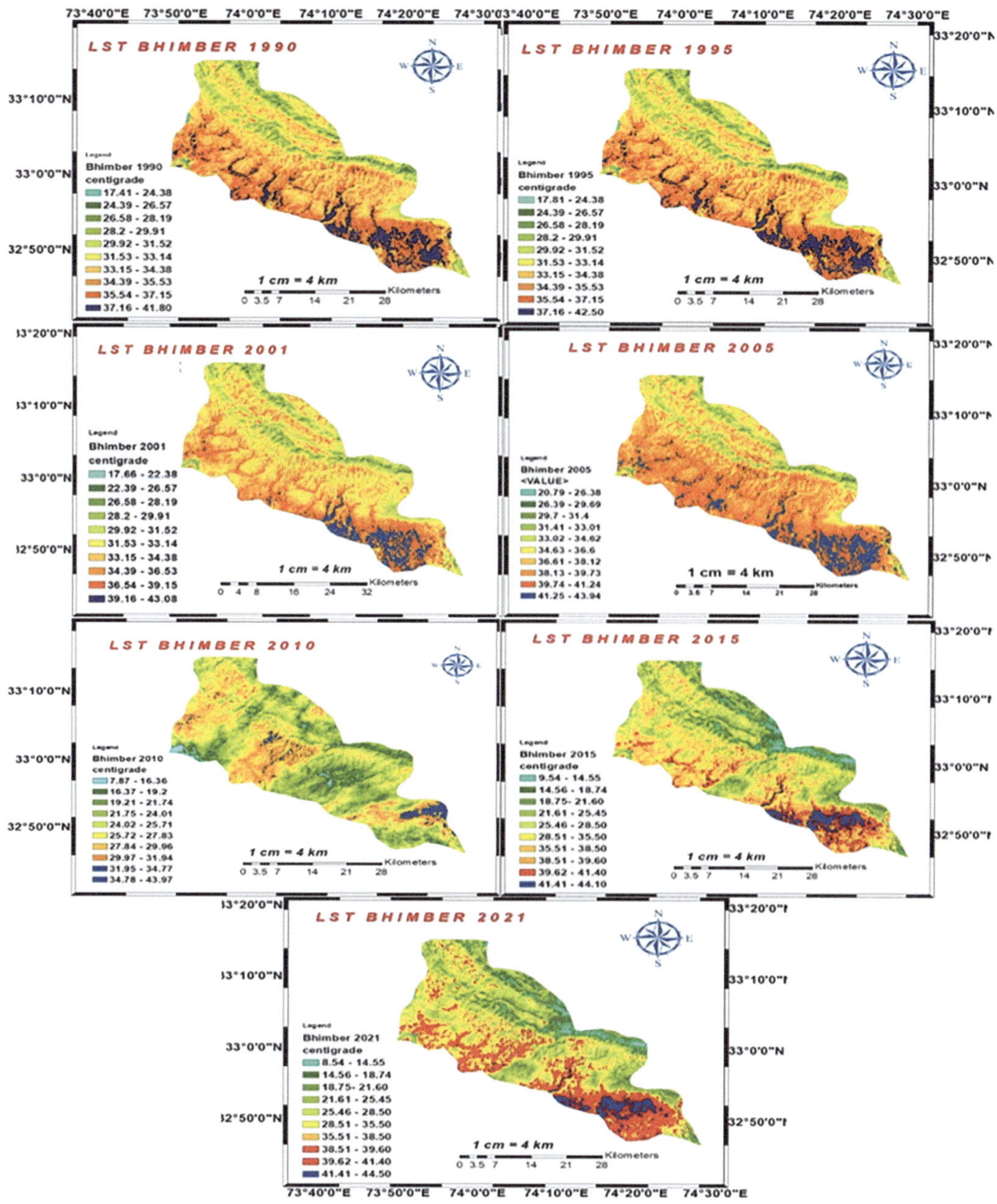

Fig. 4 Maps of LST for District Bhimber from 1990 to 2021

exposure of land surface to radiant solar energy as forests play a vital role in controlling land surface temperature (LST) by lowering albedo surfaces and high evapotranspiration and the same situation was observed by Prevedello, 2019. The role in controlling land surface temperature by forest cover was also proved by our regression analysis (R square value = 0.83) that showed a strong negative association between forest cover and land surface temperature (LST). A drastic decrease in forest cover area might be due to forest fire as forest fire effectively increases the bareness of rocks and ultimately rocks exposed. Similarly, the diverse landscape of the area is

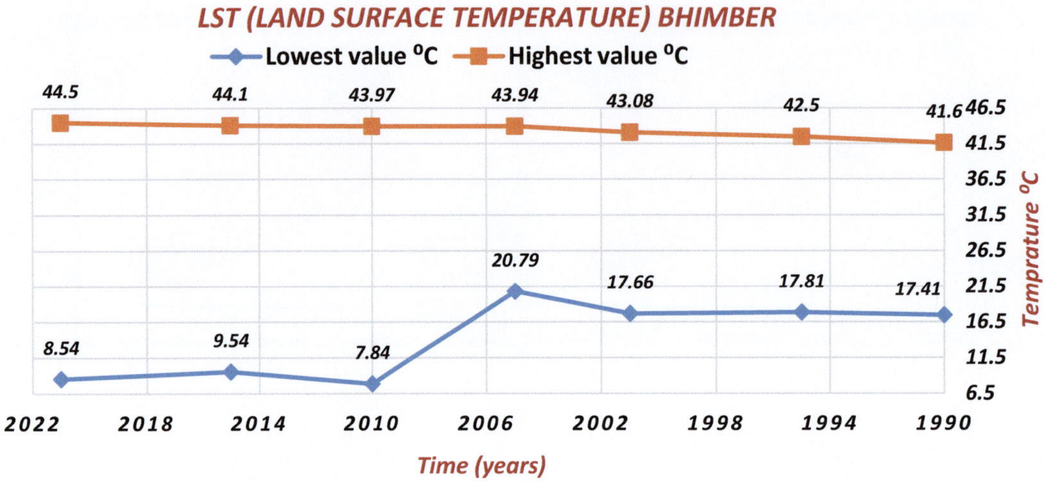

Fig. 5 Graphical representation of LST value trends from 1990 to 2021

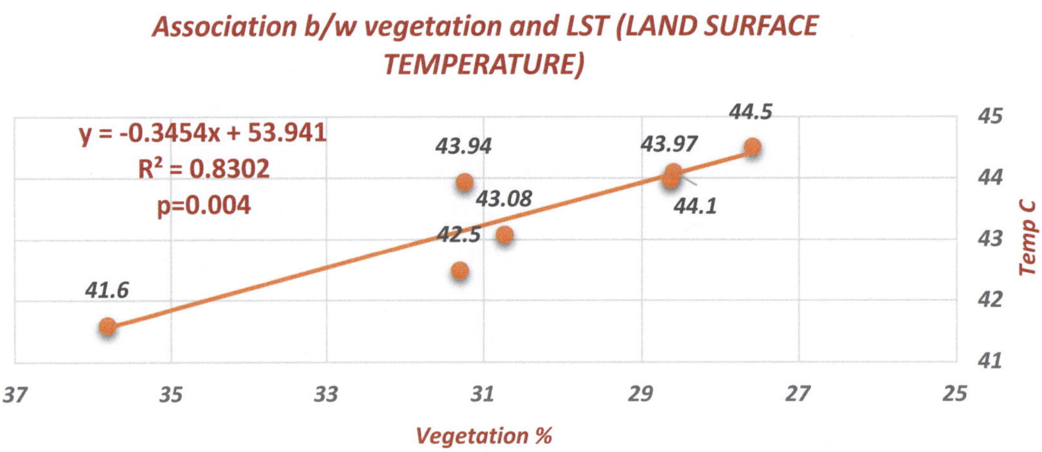

Fig. 6 Regression graph between forest cover and temperature

subject to loss of biodiversity due to more and more human concentration in the area by different activities (Khan & Chen, 2025). Our maps for LULC (Fig. 3) and results (Table 2) indicate a random trend in rock surface area due to the burning of vegetation cover by fire. Another random trend is also seen in lower temperature values (Table 4) in District Bhimber due to the presence of cold run-off water of snow-covered mountains.

The value 9.94% increase in built-up area recorded in respective study area might be due to an increase in population and migration of people toward Bhimber city due to cross-boarder fire along the Line of Control from 2001 onward. After 1995, results indicate a built-up area increased sharply. Increase in built-up area, increased albedo surface, and ultimately involved in the rise of land surface temperature (LST) (Hussain & Karuppannan, 2023). Predictions analysis indicates that Bhimber district is projected to lose 8626 hectares of forest cover and experience a temperature increase of 2.35 °C from 2021 to

Table 5 Future prediction for forest covers up to 2050 for District Bhimber

Years	Vegetation (ha)	Vegetation age (%)	Temperature (°C)
1990	48,219.9	35.82	41.6
1995	42,714.6	31.3	42.5
2001	42,539.3	30.73	43.08
2005	42,330.4	31.24	43.94
2010	38,664	28.64	43.97
2015	38,520.1	28.6	44.1
2021	37,310.1	27.6	44.5
2025	35,239.0	26.1	45.2
2030	34,477.3	25.7	45.4
2035	32,779.9	24.6	45.7
2040	31,343.4	23.5	46.0
2045	30,315.8	22.8	46.4
2050	28,684.0	21.7	46.9

2050, which will adversely affect the environment.

Conclusion

Remote sensing-based studies provide a sound monitoring for forest tracking, changes in forest cover, and resultant environmental consequences. This study is conducted to explore the forest status of Bhimber AJK that was relatively unexplored region and having very diverse types of forest cover. The dominant coniferous plant is *Pinus walichiana*. The result of this study confirmed a hazardous decrease of forest cover over the past three decades by 8.22%. The study area has lost

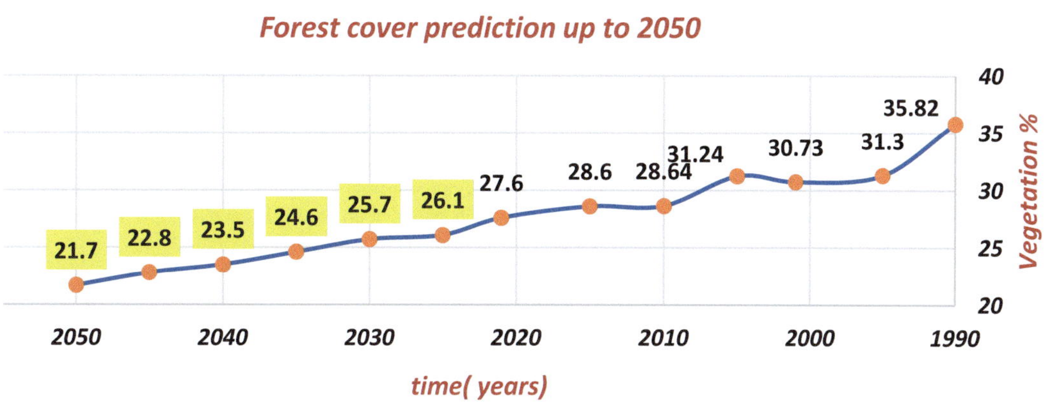

Fig. 7 Graphical presentation of prediction for District Bhimber 2050

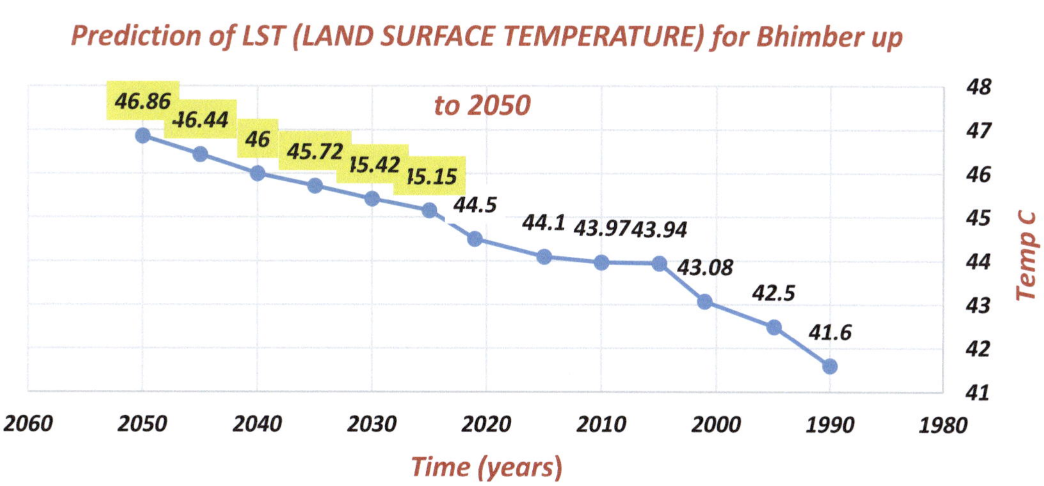

Fig. 8 LST (land surface temperature) prediction of Bhimber up to 2050

its 10,909.84(ha) area of forest cover from 1990 to 2021. This removal of forest cover ultimately contributes an increase in land surface temperature by 2.9 °C over the past three decades. Despite the presence of forest department and complete management system to safeguard forest, degradation of forest to this extent is very shocking, that signaling toward the carelessness, corruption, poor management, and knowledge about the worth of forest cover. If such a situation continued, our prediction analysis demonstrated that District Bhimber will experience a rise of 2.35 °C in land surface temperature (LST) due to the loss of 8626 ha of forest cover up to 2050, which amounts the 5.75% at the start of May. Such an increase in temperature will be harmful for crops production, water resources, and life. To avoid such hazardous situation, this study concludes an urgent need to raise awareness among local people about the importance of forest cover and provide valuable insights for policymakers to implement district-level measures in order to safeguard the forests. The following actions to safeguard forest recommended, prohibited trading, and timber mafia should be banned. There should be construction of fire breaker (furrows) after 50-m distance, plantation of eatable fruiting plants instead of non-eatable fruiting plants that will be beneficiary from both economical and forest regeneration point of views.

There is a dire need for strict policies to stop the deforestation in the area. This may include the illegal marketing and sale of forest wood for industry and fossils. On the other hand, it is suggested that the locals should follow the plant-growing campaigns in the government and educational institutes to increase the forest by planting trees.

References

Ahmad, A., Ahmad, S. R., Gilani, H., Tariq, A., Zhao, N., Aslam, R. W., & Mumtaz, F. (2021). A synthesis of spatial forest assessment studies using remote sensing data and techniques in Pakistan. *Forests, 12*(9), 1211.

Akbar, K. F. (2017). Potential impacts of climate change on plant diversity of hilly areas of Azad Kashmir and their mitigation: A review. *Journal of Mountain Area Research, 2*, 37–44.

Ali, S., Ali, H., & Raza, K. (2021). Environmental implications of deforestation: A time series study of Pakistan. *Pakistan Journal of Humanities and Social Sciences, 9*(2), 174–186.

Bokhari, T. Z., Ahmed, M., Khan, Z., Siddiqui, M. F., Zafar, M. U., & Malik, S. A. (2013). Dendroseismological potential of pine tree species of Azad Jammu and Kashmir-Pakistan: A preliminary study. *Pakistan Journal of Botany, 45*(6), 1865–1871.

Ellison, D., Morris, C. E., Locatelli, B., Sheil, D., Cohen, J., Murdiyarso, D., Gutierrez, V., Noordwijk, M. V., Creed, I. F., Pokorny, J., Gaveau, D., Spracklen, D. V., Tobella, A. B., Ilstedt, U., Teuling, A. J., Gebrehiwot, S. G., Sands, D. C., Muys, B., Verbist, B., & Sullivan, C. A. (2017). Trees, forests and water: Cool insights for a hot world. *Global Environmental Change, 43*, 51–61. https://doi.org/10.1016/j.gloenvcha.2017.01.002

Hansen, M. C., Roy, D. P., Lindquist, E., Adusei, B., Justice, C. O., & Altstatt, A. (2008). *A method for integrating MODIS and Landsat data for systematic monitoring of forest cover*.

Huang, B., Wang, J., Song, H., Fu, D., & Wong, K. (2013). Generating high spatiotemporal resolution land surface temperature for urban heat island monitoring. *IEEE Geoscience and Remote Sensing Letters, 10*(5), 1011–1015.

Hussain, S., & Karuppannan, S. (2023). Land use/land cover changes and their impact on land surface temperature using remote sensing technique in district Khanewal, Punjab Pakistan. *Geology, Ecology, and Landscapes, 7*(1), 46–58.

Khalid, F., Taj, M. B., Jamil, A., Kamal, H., Afzal, T., Iqbal, M. J., & Tirmizi, S. A. (2020). Deforestation dynamics in Pakistan: A critical review: Deforestation dynamics. *Proceedings of the Pakistan Academy of Sciences: B. Life and Environmental Sciences, 57*(3), 27–34.

Khalid, N., Ullah, S., Ahmad, S. S., Ali, A., & Chishtie, F. (2019). A remotely sensed tracking of forest cover and associated temperature change in Margalla hills. *International Journal of Digital Earth, 12*(10), 1133–1150.

Khan, M., & Chen, R. (2025). Assessing the impact of land use and land cover change on environmental parameters in Khyber Pakhtunkhwa, Pakistan: A comprehensive study and future projections. *Remote Sensing, 17*(1), 170.

Maillard, O., Vides-Almonacid, R., Salazar, Á., & Larrea-Alcazar, D. M. (2023). Effect of deforestation on land surface temperature in the Chiquitania Region, Bolivia. *Land, 12*(1), 2.

Maus, V., Giljum, S., Gutschlhofer, J., da Silva, D. M., Probst, M., Gass, S. L., & McCallum, I. (2020). A global-scale data set of mining areas. *Scientific Data, 7*(1), 1–13.

Mishra, M., Santos, C. A. G., do Nascimento, T. V. M., Dash, M. K., da Silva, R. M., Kar, D., & Acharyya, T. (2022). Mining impacts on forest cover change in a tropical forest using remote sensing and spatial information from 2001–2019: A case study of Odisha

(India). *Journal of Environmental Management, 302*, 114067.

Mohammad, N., Wajid, R., Sultan, H., Khan, N., Khurshid, M., Ali, W., & Bohnett, E. (2024). Occurrence, probable causes, and management of forest wildfires in the Northern Highlands of Pakistan. *Environmental Challenges, 15*, 100930. https://doi.org/10.1016/j.envc.2024.100930

Pérez-Rey, I., Sarro, R., Tomás, R., Alejano, L. R., Gutiérrez, L. H., Mateos, R. M., & Riquelme, A. (2023). A brief review of the effect of wildfires on rockfall occurrence. *IOP Conference Series: Earth and Environmental Science, 1124*(1), 012122.

Prevedello, J. A., Winck, G. R., Weber, M. M., Nichols, E., & Sinervo, B. (2019). Impacts of forestation and deforestation on local temperature across the globe. *PLoS One, 14*(3), e0213368.

Rahardja, D. (2021). Statistical time-series forecast via microsoft excel (FORECAST. ETS) built-in function. *Journal of Research in Applied Mathematics, 7*(11), 69–73.

Ramzan, M., Saqib, Z. A., Hussain, E., Khan, J. A., Nazir, A., Dasti, M. Y. S., & Niazi, N. K. (2022). Remote sensing-based prediction of temporal changes in land surface temperature and land use-land cover (LULC) in urban environments. *Land, 11*(9), 1610.

Rich, B. (2014). *Mortgaging the earth: World bank, environmental impoverishment and the crisis of development*. Routledge.

Richard, J. U., & Abah, I. A. (2019). Derivation of land surface temperature (LST (land surface temperature)) from LANDSAT 7 & 8 imageries and its relationship with two vegetation indices (NDVI and GNDVI). *International Journal of Research, 7*(2), 108–120.

Shafiq, M., & Mahmood, S. (2022a). Spatial assessment of forest cover change in Azad Kashmir, Pakistan. *Advanced GIS, 2*(2), 62–69.

Shafiq, M., & Mahmood, S. (2022b). Spatial assessment of forest cover change in Azad Kashmir, Pakistan. *Advance GIS, 2*(1), 689. Retrieved from https://publish.mersin.edu.tr/index.php/agis/article/view/689

Suryabhagavan, K. V., Alemu, M., & Balakrishnan, M. (2016). GIS-based multi-criteria decision analysis for forest fire susceptibility mapping: a case study in Harenna forest, southwestern Ethiopia. *Tropical Ecology, 57*(1), 33–43.

Tasmia, M., Noreen, S., & Ali, S. S. (2014). Causes and effects of deforestation on Union Council Ashkot, Neelum Azad Jammu and Kashmir: a community perspective. *Middle-East Journal of Scientific Research, 22*(6), 933–936.

Younis, S. M. Z., & Iqbal, J. (2015). Estimation of soil moisture using multispectral and FTIR techniques. *The Egyptian Journal of Remote Sensing and Space Science, 18*(2), 151–161.

The Role of Plant Biodiversity in Blue Infrastructure Sustainability

Shumaila Anam, Muhammad Majeed, Laiba, Maida Mobeen, Jamila Naz, Awais Amin, Muhammad Atiq Ashraf, Waseem Ahmed Khattak, and Muhammad Anas

Abstract

In response to the compounding challenges of urbanization and climate change, urban planners are increasingly recognizing the importance of incorporating blue infrastructure into their designs to promote more resilient and adaptable cities. By incorporating diverse plant species, blue infrastructure can be transformed into resilient ecosystems that provide numerous benefits, including habitat provision, water purification, and stormwater management. The strategic integration of plant biodiversity into blue infrastructure is vital for enhancing the ecological, social, and economic benefits of urban environments, and promotes multifunctional benefits in urban ecosystems. Central to the effectiveness of blue infrastructure is plant biodiversity, which contributes significantly to the functionality and sustainability of these systems. Plant biodiversity is important for assuring the environmental uprightness of blue infrastructure. The incorporation of diverse plant species into aquatic ecosystems can significantly enhance their resilience and ecological functioning, including improved nutrient cycling, reduced erosion, and enhanced soil stability. The diverse range of plant species in urban ecosystems supports a variety of ecosystem services, including carbon sequestration, air quality improvement, and the creation of recreational areas that are critical for maintaining the health, well-being, and quality of life of urban populations. This chapter explores the vital role of native plant species in supporting local biodiversity, enhancing ecosystem resilience, and mitigating the impacts of climate variabil-

S. Anam
Department of Botany, University of Agriculture, Faisalabad, Punjab, Pakistan

M. Majeed (✉) · M. Mobeen · A. Amin
Department of Botany, University of Gujrat, Gujrat, Punjab, Pakistan
e-mail: m.majeed@uog.edu.pk; 23011706-011@uog.edu.pk

Laiba
Department of Orthotics and prosthetics, Government College University, Faisalabad, Punjab, Pakistan

J. Naz
Department of Forestry and Range Management, Bahauddin Zakariya University, Multan, Punjab, Pakistan

M. A. Ashraf
National Key Laboratory for Germplasm Innovation and Utilization of Horticultural Crops, Huazhong Agricultural University, Wuhan, Hubei, China
e-mail: atiqashraf25@webmail.hzau.edu.cn

W. A. Khattak · M. Anas
Faculty of Biological Sciences, Department of Plant Sciences, Quaid-I-Azam University, Islamabad, Pakistan
e-mail: waseemkhattack@bs.qau.edu.pk; muhammadanas@bs.qau.edu.pk

M. M. Anwar, A. Aziz (eds.), *Sustainable Landscape for Urban Resilience*, Sustainable Development Goals Series, https://doi.org/10.1007/978-3-031-99836-2_10

ity. While, by incorporating native plants into urban green spaces, cities can create inviting, sustainable areas promote mental well-being, community interaction, and environmental stewardship. Additionally, incorporating it into educational elements, these green spaces can raise awareness about biodiversity, ecosystem health, and environmental sustainability, inspiring urban residents to adopt eco-friendly practices. Urban areas can create thriving ecosystems by supporting diverse plant life that attract pollinators, beneficial organisms, and other wildlife, ultimately enriching the ecological fabric of cities. The renewability of blue infrastructure is personally connected to the management practices hired in nonrural planning. Furthermore, by conserving and restoring native plant species, cities can foster more resilient ecosystems, better equipped to mitigate the effects of climate change and support the well-being of urban residents. Incorporating native plants into bioswales and rain gardens can effectively manage stormwater runoff, decrease maintenance needs, and minimize resource consumption. Green corridors and similar initiatives can serve as powerful tools for promoting biodiversity awareness, engaging local communities in conservation efforts, and inspiring collective action.

Keywords

Biodiversity · Ecological benefits · Environment · Rainwater · Management · Ecological diversity · Eco-infrastructure · Wildlife · Species Preservation · Green · Plants

Introduction

Investigating the importance of plant biodiversity in blue infrastructure sustainability can inform innovative solutions for urban ecosystem management, enhancing both environmental resilience and human well-being (Liu et al., 2021). Blue infrastructure, which consist of water bodies and associated ecosystems, profits properly from the activeness of various plant species.

This relationship increases different environment services and consists of aquatic health enhancement, habitat provision, and climate resilience. Blue–green infrastructure (BGI) encompasses a range of ecological engineering and green infrastructure strategies that enhance the health and resilience of terrestrial, freshwater, and marine ecosystems (Drosou et al., 2019). Cities' blue–green infrastructure (BGI) incorporates natural capital elements to ensure long-term sustainability, yielding numerous benefits for ecosystems, communities, businesses, and cultural heritage. This study examines the benefits of blue–green infrastructure (BGI) by addressing the research question: Does the implementation of BGI contribute to improved habitat connectivity and increased biodiversity in urban environments? To investigate this, the BEETLE tool set was integrated with the LUCI framework and applied to Christchurch, New Zealand, providing new insights into the city's ecological dynamics (Ali et al., 2023; Arif et al., 2021; Arshad et al., 2022; Aziz, Anwar, Majeed, Albanai, et al., 2023; Nguyen, 2022). Three species were chosen for their ecological representatives, and their parameterization was informed by a combination of ecological theory and expert insight. The introduction of BGI led to a significant expansion of habitats for kereru (3.3%) and paradise shelduck (2.5%), thereby improving habitat connectivity. We propose exploring opportunities to establish additional native patches around the catchment, aiming to meet the minimum 10% indigenous cover target. While BGI is a valuable conservation strategy, it has limitations, and the restoration of threatened wildlife to the city will require additional measures, including predator-fenced. The analysis of socio-eco-spatial connectivity in this research was structured around four interconnected dimensions.

Currently, urban areas are home to more than 50% of the global population, and this number is steadily rising. Consequently, vast expanses of natural habitats worldwide will be transformed into human-dominated landscapes. The expansion of urban areas is causing the degradation and fragmentation of ecosystems situated at the intersection of hills, harbors, plains, rivers, and estuaries, leading to a decline in biodiversity. Crucial biological processes, including reproduction, migration, and resource allocation, are severely impacted (Francis & Chadwick, 2013). Due to the diverse responses of species to landscape patterns, patch composition, and cultural landscape changes, a species-based approach is required to examine the functional impacts of landscape change, as evident in case studies from the United Kingdom, China, and Mexico. There is a growing acknowledgment of the significance of urban environments as habitats for wildlife and their intersections with human populations.

Background

The multidimensional nature of biodiversity encompasses a wide range of elements, including taxonomic, functional, phylogenetic, genetic, and landscape diversity, among others. Research on biodiversity's contribution to ecosystem functions and services frequently neglects to account for its inherent multidimensionality. As a result of this oversight, our present comprehension of the ecological and environmental implications of biodiversity decline is mainly restricted to the insights gained from unidimensional research.

The urbanization process, marked by the transition from vegetated and permeable surfaces to impermeable and heat-absorbing materials (asphalt, concrete, brick) (Baliello, 2020), results in warmer urban microclimates, increased storm runoff, and reduced biodiversity due to habitat destruction. In various urban centers, poor air quality resulting from local emissions of NOx, ozone, and particulate matter has emerged as a significant public health concern. Projected climate change, characterized by increased frequency and severity of extreme weather events (such as heat waves and intense storms), is expected to worsen urban issues like stormwater runoff, pollution, and heat stress. As a result, vegetation has emerged as a critical component of urban landscapes, with its role evolving rapidly in recent years. As the global population becomes increasingly urbanized (Seto et al., 2010), urban vegetation is anticipated to play a multifaceted role that extends far beyond its traditional aesthetic and culinary purposes. Society's increasing recognition of the adverse effects of urbanization, coupled with the understanding of vegetation's benefits, has led to these demands.

Green infrastructure, comprising natural and engineered elements like parks, green roofs, and walls, is crucial for reducing the environmental footprint of cities and addressing climate change challenges. Municipal and local governments are progressively incorporating green infrastructure into their urban planning stratcgics and policies, driven by the growing awareness of its benefits. The capacity of green infrastructure to provide ecosystem services, including stormwater management, temperature regulation, biodiversity conservation, and capturing air pollutants, is essential to select the most suitable plant species or combinations of species (Coutts & Hahn, 2015). As a result, urban plantings that fail to meet expectations can have serious repercussions for urban sustainability, community health, and environmental quality.

Scope

Integrating Nature-Based Solutions in Urban Planning

Nature-based solutions have become a mainstream approach to addressing urban environmental challenges, yielding numerous economic, social, and ecological advantages. BGI provides a unified planning framework for creating urban environments that balance human well-being, environmental sustainability, and economic viability. BGI utilizes integrated networks of natural and engineered landscape elements to mitigate urban flood risk. The advantages of BGI include urban cooling, water management, recreation, eco-friendly design, and habitat creation, contributing to more sustainable, resilient, and livable cities. For example, apart from recreational and protected green space, backyards, rain gardens along roadsides, and playing field perimeters offer general wildlife habitats at a large to fine scale. In addition to formal green spaces, urban areas feature various habitats that support local wildlife, including residential backyards, roadside rain gardens, and sports field buffers. Green roofs and living walls have provided habitats for lizards in cities if niche modifications are applied to make them more lizard friendly and predator protected (Aziz, Anwar, Majeed, Fatima, et al., 2023; Bashir et al., 2023; Bhattacharya et al., 2023; Haq, Tariq, et al., 2022; Haq, Yaqoob, et al., 2022; Nguyen et al., 2021). Water birds are attracted to constructed urban wetlands, including basins and flood-way lawns, designed for flood control, public amenity, and other environmental benefits.

Artificial urban wetlands, comprising basins and flood-way lawns, have been found to attract water birds, in addition to delivering benefits such as flood control, public recreation, and ecological enrichment. To optimize wildlife movement corridors and conserve meta-populations, researchers have utilized least-cost models, identifying preferred routes in natural environments. BEETLE, a landscape ecology tool developed by Forest Research (UK), is applied in this context to assess habitat connectivity. A geographic information system (GIS) is utilized to spatially capture the characteristics and requirements of a diverse range of ecologically representative focal species. To enhance efficiency, the BEETLE workflow has been arithmetically automated and embedded within the (LUCI) framework. By utilizing spatially explicit methods, LUCI facilitates the examination of a landscape's capability to provide a range of ecosystem services, including flood mitigation, habitat connectivity, erosion control, carbon storage, water quality enhancement, and trade-off analysis (Trodahl et al., 2017). With its intuitive interface, the habitat connectivity tool is readily accessible to wildlife managers seeking to address connectivity issues, assuming known habitat needs and dispersal traits. This study marks the first application of the tool in New Zealand, where it has been tailored to accommodate the unique habitat needs and behaviors of the country's endemic wildlife (Fig. 1).

Limitations

Factors Leading to Plant Failure

Urban plantings may decline or not meet expected results in open environments for a variety of reasons, including suboptimal soil conditions, adverse environmental conditions (such as extended drought in nonirrigated plantings), low quality of planting stock and techniques, poor location, vandalism, pests and pathogens, and poor plant choice (Hassan, Haq, Ahmad, et al., 2022; Hassan, Haq, Majeed, et al., 2022; Hussain et al., 2023; Lu et al., 2011; Majeed et al., 2020; Majeed, Bhatti, Pieroni, et al., 2021; Majeed, Tariq, Anwar, et al., 2021). Urban landscapes are often considered hostile environments for vegetation, with many factors that can simultaneously impede plant growth.

Potential of Urban Landscapes

However, because the urban landscape is a highly varied environment that offers a range of microsites, niches, and settings that are appropriate for various kinds of vegetation and green infrastructure, it is also feasible to appreciate it more deeply. According to this constructivist viewpoint, there is a great deal of space for matching plant types to various environments, which will boost species variety and create adaptive urban landscapes that can both adapt to and lessen the consequences of climate change (Diamond, 1988).

Approaches to Effective Plant Selection

More conventionally, it involves matching up suitable plant species with specific site conditions and resource availabilities plus aesthetic, cultural, and functional requirements. This approach can be highly demanding, and the difficulty is increased by the limited availability of biological tolerances for most plant taxa as described in Table 1. Furthermore, due to the nature of the demonstration of present-day urban vegetation informing leaders choices on which species should be used, one is likely to end up with a relatively few numbers of what is more of

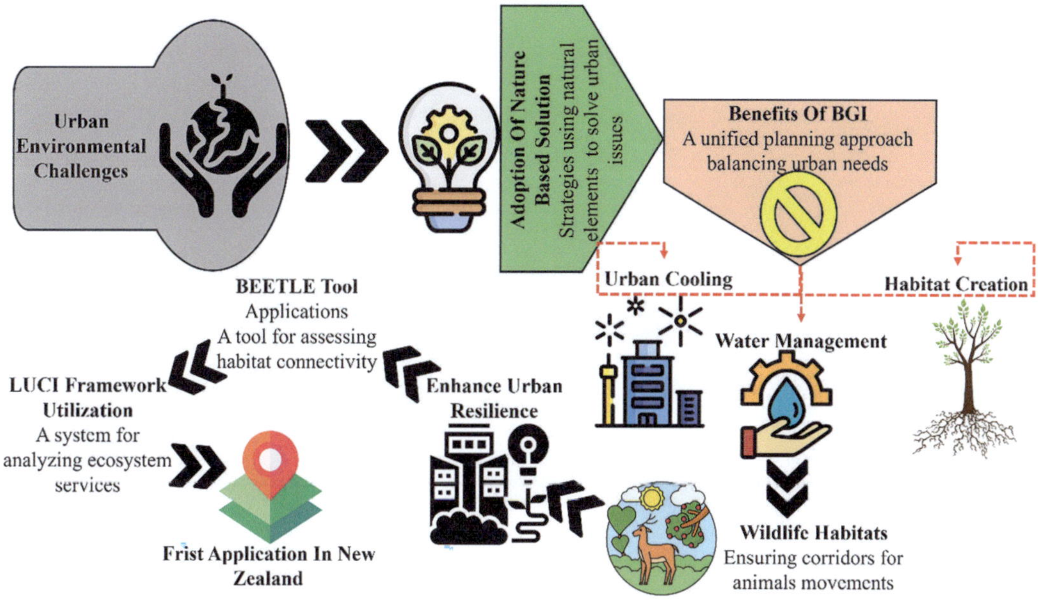

Fig. 1 Integration of nature-based solutions in urban planning

Table 1 Approaches, description, and benefits to face the environmental challenges

Approach	Description	Benefits	Example plants
Native plant selection	Select plants that are indigenous to the area's ecology	Enhances drought resistance, biodiversity, and habitat creation	Purple coneflower, black-eyed Susan
Water-tolerant species	Choose plants that thrive in damp conditions or fluctuating levels of water	Supports stormwater management, flood control, and resilience	Water lilies, reed canary grass
Drought-resilient plants	Choose species that have the ability to endure prolonged dry conditions	Reduces demand of water, especially in arid areas	Sedum, agave, lavender
Ecological compatibility	Ensure that selection of plant aligns with local fauna and habitats	Supports pollinators, promotes ecological balance, and reduces invasive species	Milkweed, goldenrod, willow trees

"extreme species" being adopted to plant landscapes in a regional and global sense (Jamil et al., 2022; Khan et al., 2022; Khoja et al., 2022; Majeed et al., 2020; Majeed, Bhatti, & Amjad, 2021; Roman et al., 2018). Then existing species will also find it very challenging to adapt to hotter and drier climate conditions (Fig. 2).

The Need for Improved Selection Processes

The selection of species becomes much more difficult because green foundations might have several goals, such as improving biodiversity, cooling, managing stormwater, and removing pollutants. Plant selection processes may be optimized by matching plant traits to site conditions and intended outcomes; however, this requires a clear methodology and approach, which can be difficult to implement across interdisciplinary teams. It has been demonstrated that plant determination procedures for trees in urban settings are useful, and they have even been used to help create reports for the introduction of trees in urban settlements all over the world. To aid in the development of final plant reports, for example, the design purpose should be clearly defined in

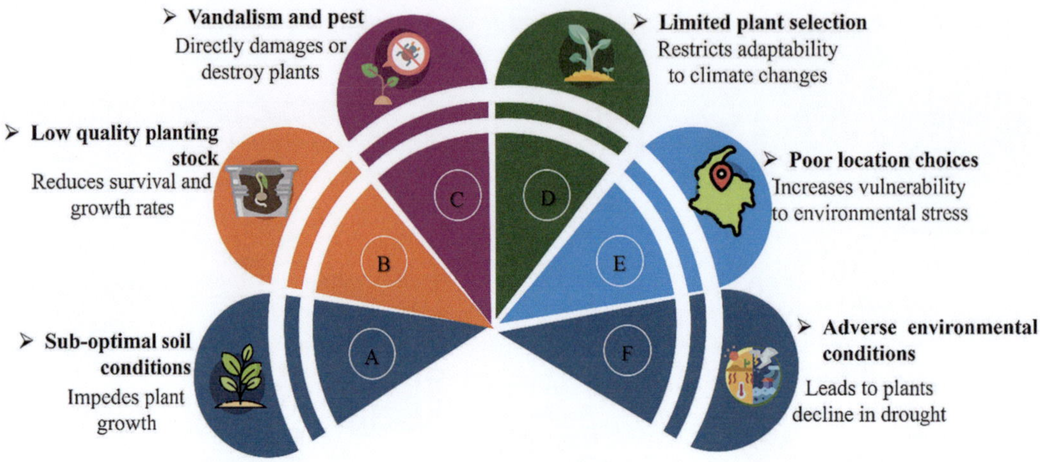

Fig. 2 Urban planting failure due to the environmental challenges

order to produce criteria for making decisions and ranking species according to location (ecological or social demands), economic, and social variables. With a focus on the growth of plants in site situations where soil quantities are fairly limited, this model has since been extended to include the explicit picking of tree species in green infrastructure (Gardner et al., 1999; Majeed, Khan, et al., 2022; Majeed, Lu, et al., 2022; Majeed, Lu, Anwar, et al., 2023; Majeed, Tariq, Anwar, et al., 2021). With the wide range of synthetic green technologies available, selection tactics should not be restricted to just choosing each tree for urban settings. Furthermore, such created approaches need to take into account future situations and their consequences with regard to the deployment of green infrastructure and plant resistance. This is because the several disciplines that make up green infrastructure make the processes for selecting plants strong and sufficiently explicit to create diverse and adaptable urban landscapes.

Integrating Ecological Approaches for Resilience

We then go over how nature-based methods could be used to choose suitable plants for green infrastructure that would withstand current and upcoming environmental circumstances. In addition to offering social services like education, leisure, social capital, and economic development, green infrastructure can also improve human health and well-being (Coutts & Hahn, 2015). However, our strategy focused on choosing resilient and effective plants. When evaluating the potential of various species, it is important to better incorporate plant science, according to earlier assessments on plant selection in green infrastructure.

Addressing Knowledge Gaps in Plant Performance

We used earlier methods that make species selection based on availability, fashion, and social requirements to implement another plant choice system that includes the transport of environment support resistance of establishment conditions in green foundation. This is because, although there is information available about the resilience of various plant species to urban conditions, data regarding the contributions made by various species in stormwater management, temperature regulation, biodiversity promotion, and emission of harmful air pollutants is largely nonexistent and difficult to measure. To illustrate the relationship between plant traits and native species distribution in connection with resilience as well as the provision of benefits such as managing stormwater, thermal regulation, biodiversity improve-

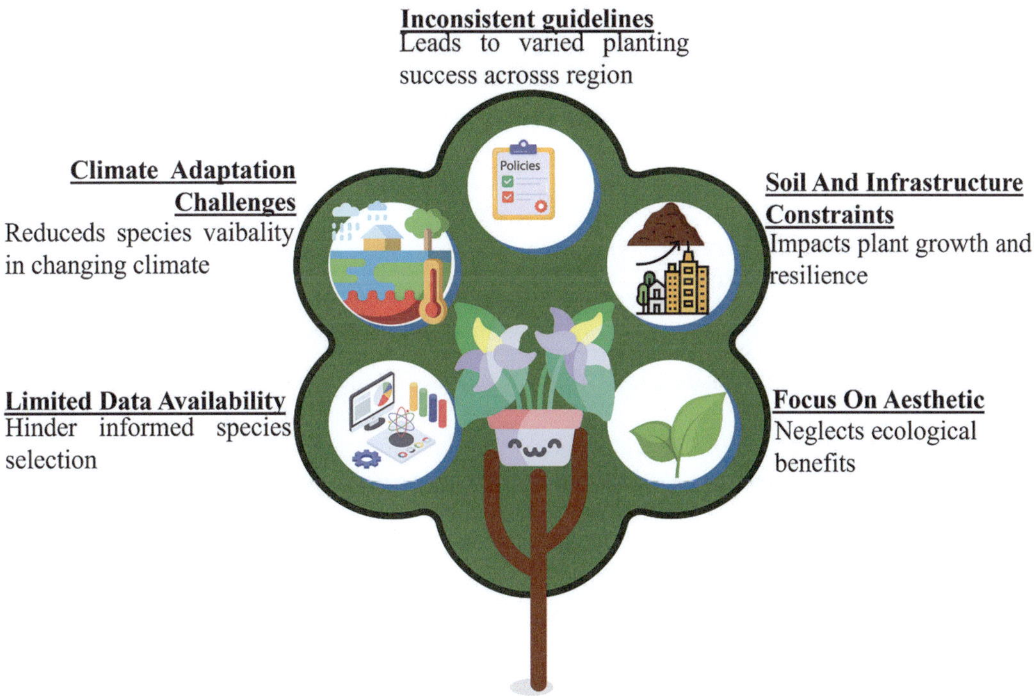

Fig. 3 Ineffective plant species selection for urban resilience

ment, and quality of air improvement, we draw on recent studies that have assessed vegetation over green infrastructure and urban purposes (Gamon et al., 2019; Khan et al., 2024; Majeed, Muhammad, Hussain, et al., 2023; Nasar, Muhammad, et al., 2024; Nasar, Rasool, et al., 2024; Waheed et al., 2022). We organized our literature using cities from around the world and did not limit our guidelines to just one climate type because the conditions for planting or location details in the green framework are frequently different from the surrounding conditions due to soil limitations or the impact of urban infrastructure on microclimate. As an illustration of this strategy, which will be especially relevant for cities experiencing drier and warmer climates or periods of episodic drought, we concentrated on resilience to two major climate change stressors in our study: heat and severe drought. Intensity and dry spell pressure are crucial in a changing environment, and there are definitely established correlations between plant traits and responses to these stresses. This is by no means an exhaustive review (Fig. 3).

Case Study Application of the Selection Framework

In order to illustrate how the plant selection process might be used in new plantings into Australian urban areas, we conclude with a case study. We also discuss the limitations of these approaches and make a list of potential lines of inquiry for further study. Finally, we consider how plant diversity might be utilized to achieve a range of planting goals while lowering the likelihood of failure in urban environments (McClean et al., 2005).

Addressing Urban Challenges with Green Stormwater Infrastructure

Managing Urban Flooding and Water Pollution

Monitoring of water pollution and town flooding for many metropolitan locations, flooding, structural deterioration, and water source contamination are major problems. Green stormwater infrastructure (GSI) is being implemented by more and more communities to solve these issues (Heidari et al., 2023).

The Role and Design of Green Stormwater Infrastructure

The GI structure, which is basically stormwater management, includes the GSI category. It can take various forms, including green roofs, bioswales, and rain gardens. Few GSI are equally well-suited for stormwater management, and when deciding how to best achieve GSI in urbanized settings, developers should take into consideration both the site's exceptional qualities and the biological traits and limitations of each type. In a similar vein, practitioners ought to investigate the GSI types that clients like. Because designers have limited information regarding the extent to which people prefer one type of GSI over another, this is a test. It is easy to imagine that two GSI situations that are equally suited for monitoring stormwater could have basically distinct inclination values (Boley, 2009).

User Preferences and the Impact on GSI Design

Although there is some evidence that individuals prefer to live in areas with a lot natural plants and trees, nothing is known about how people feel about GSI. We believe that different sorts of urban greenery are viewed differently than GSI, which can take many diverse shapes and planting schemes. Perennial vegetation, long grasses, and other native plants that are abundant in biodiversity are frequently included in GSI; yet, these assemblages may be perceived as disorganized. Designers and planners run the risk of creating surroundings that people dislike if they do not understand why individuals prefer particular types of GSI (Sennett & Sendra, 2020). People are far less inclined to invest their time in situations they do not enjoy since they do not think they are good.

Understanding Designer Vs. Public Preferences

Furthermore, it is unclear if non designer-trained designers' tastes for creating GSI are comparable to those of nondesigners. According to research, planners, structural engineers, and landscape architects have distinct tastes in parks and other landscapes than do laypeople. Understanding the distinctions between the preferences of professionals and the general public is necessary because GSI implementation in cities is a requirement for professionals. Filling up this knowledge gap will promote GSI organizations that are valued and possibly even treasured by the public and master designers.

The Importance of Green and Blue Infrastructure in Urban Environments

The Concept and Benefits of Green and Blue Infrastructure

Natural cycles can be restored and vital urban functions in cities can be enhanced by reintroducing and reintegrating natural elements into constructed settings. Both human well-being and ecological integrity benefit from such actions. The GI framework, which is based on the fundamental ideas of contentedness and multi functionality, is a planned and methodical approach to integrating natural resources, such as vegetation and wetlands into the urban environment. It is also recognized as an NbS for cities looking to boost climate resilience and enhance ES (Hobbie & Grimm, 2020; Majeed, Muhammad, Nawaz, et al., 2023; Muhammad et al., 2023; Muhammad, Basit, Arooj, et al., 2024; Sinha et al., 2023; Tariq et al., 2023; Tassadduq et al., 2022; Ullah et al., 2022). The aquatic environments and their functions are incorporated into this framework by the GBI concept, also referred to as BGI. Researchers and professionals in the fields of hydrological and engineering studies, as well as those working in the scope of climate adaptation and resilience research, are the primary users of the concept.

This article uses the phrase "green and blue foundation" (GBI) since it stresses water management practices that are relevant to metropolitan areas in the majority of developing countries, even though the funding under analysis contains comparative terms.

Global Applications and Case Studies of GBI

The literature does a fantastic job of explaining GBI. For their various environmental administrations, city experts are using GBI procedures, such as controlling the surrounding environment (e.g., Stuttgart, Germany; Melbourne, Australia); reducing surface overflow to Rotterdam, the Netherlands, and protecting water assets and capturing carbon in Durban, South Africa. A variety of nature-related goals are linked to more practical human-oriented goals, such as quality of life (Greco & Bencardino, 2014), environmental and human health, and local job prospects, as a result of the development of GBI mediation in the city settings.

Socioeconomic Considerations in Implementing GBI

Planners and practitioners in developing nations are frequently distracted by fundamental economic issues, despite the fact that urban specialists and experts generally have a positive opinion of GBIs as a way to improve availability of nature as well as ecosystems. These include unequal mobility across different regions, substandard living conditions, and limited access to essential public services like energy and clean water. Urban engineers and technologists are unaware of green-built infrastructure (GBI) systems, despite the fact that the regulations and structures developed are intended to protect and enhance GBI deployment (Mjimba, 2019). However, the vast majority of marginalized communities still rely on ecosystem services and natural resources for their survival, means of subsistence, and access to necessities in the modern world. Through creative policy tools, a number of cities have recently learnt that incorporating green as well as blue infrastructure frameworks in land-use planning and management can enhance ecosystem services in urban areas while also improving the economic circumstances of nearby communities. In conjunction with resilience and adaptation strategies for urban areas, the green and blue framework methodology is regarded by numerous international initiatives and organizations, including the European Commission (Francis & Chadwick, 2013), the Intergovernmental Scientific and Policy Platform on Ecosystem Services and Biodiversity (IPBES), as well as the Intergovernmental Panel on Climate Change (IPCC), as a significant form of nature-based solution pertinent to urban areas.

The Importance of Urban Green and Blue Infrastructure in Developing Countries

Urbanization and Environmental Challenges

Current and future research in innovation and urban GBI is important in developing countries to get knowledge about cities; the population in these countries is about two-thirds of the world. Middle-income countries have growing urban center on which various environmental issues of future depend such as climate change. Moreover, over the next decades it is expected that human population in cities, especially in Asia and Africa, will be further concentrated and increased in numbers, especially in Africa and Asia. Socioeconomic challenges in Latin America are the result of rapid urbanization during the period of 1960s–1980s, but urban centers and cities both will be changed due to additional urbanization in future. For instance, social challenges and urban sprawl into conservation areas are the result of growth of informal settlements in urban fringes. Furthermore, climate change adaptation is the great concern of local government in developing countries, lack of resources and competing priorities, haphazard urban development gives the vulnerability. Integration of perspective of the Global South addressed global environmental problems (Najam, 2020) such as inequalities of socio-environment is reduced by this knowledge, which is aggravated by biodiversity loss and climate change through which the whole planet is effected, particularly most sectors of human society and vulnerable natural ecosystems.

Nature-Inspired Solutions for Urban Sustainability

The functionality of spatial cohesion and quality and quantity of natural areas are degraded by increasing urbanization, which is essential for ecosystem functioning and biodiversity worldwide. Movability of wildlife and existing habitat ranges are threatened by erosion of green (i.e., terrestrial) and blue (i.e., aquatic) and building activity of uncontrolled pace. Emerging mitigation solutions are due to nature-inspired engineered blue-green infrastructure (BGI), which is local scale measurement. Existing ecological infrastructure of biodiversity has interesting synergies that are offered by instruments such as improved human livability in cities, adaptation to climate change, and planned to promote sustainable storm water management. Globally declining amphibians are especially appealed by BGI, and urbanization makes it sensitive to diverse and rich vertebrate assemblage. Core of ecological corridors and habitats are improved by multispecies and multiscale protection in the Swiss lowlands (Ashrafzadeh et al., 2020); for this purpose, we integrated circuit theory-based connectivity models, ensemble models of habitat suitability, and biological and highly resolved urban–rural land-cover data. Considering a broad spectrum of amphibian biodiversity, we identified distributions of amphibian biodiversity hotspots and four landscape elements and distribution of amphibian biodiversity hotspots are identified by considering a broad spectrum of amphibian biodiversity that is essential to movability of amphibians at the regional scale, namely

(a) Wet-forest habitat.
(b) Forest edges.
(c) Riparian zones.
(d) Soil with variable moistures.

Connectivity of landscape habitat can make a substantial contribution (e.g., up to 15% of urban space in the study area) by cities that is shown in our work. We highlight that strategic locations have the importance of BGI planning locally to promote availability and permeability of “stepping stone” across urban and peri-urban areas, ecosystem functioning, and enhancing biodiversity and maintenance of connectivity of regional habitat are essential for densely populated landscape (Fig. 4).

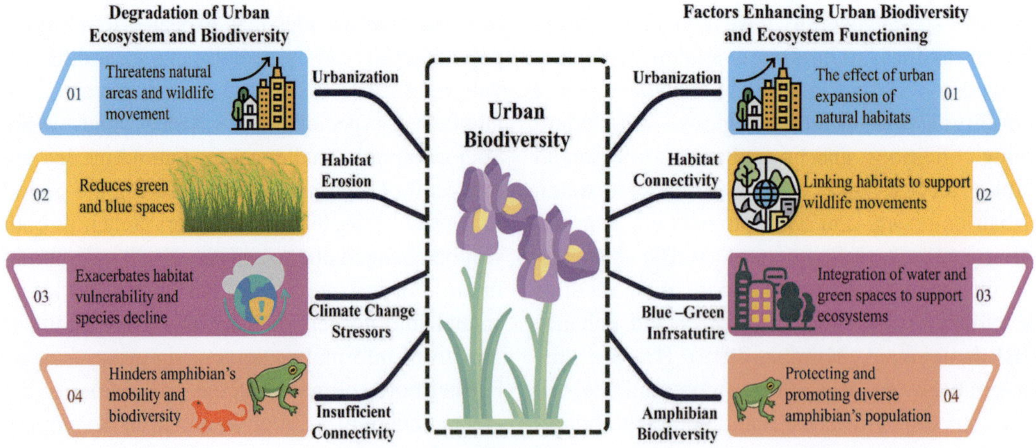

Fig. 4 Urban biodiversity faces challenges from urbanization, habitat erosion, and climate stressors, leading to species decline and reduced connectivity

However, strategic integration of green and blue infrastructure, habitat connectivity, and amphibian conservation can enhance ecosystem resilience. Balancing urban development with ecological strategies is key to fostering thriving urban biodiversity.

Addressing Socioeconomic Inequalities Through GBI

While countries that are related to the Southern Hemisphere do not contain all the Global South, and socio-economic characteristics are not the same in all these regions, economic informality and rapid haphazard urban growth that are related to common challenges and vulnerabilities are shared by some aggravated by the impacts of globalization and colonialist urban legacies. Large inequalities are demonstrated by the assessment of GDP per capita and assessment of development data that are not detailed in the global report between and within regions and countries. The literature has established trends and analytical common ground; the Global South 1 (Hussain et al., 2024; Majeed, Gulshan, et al., 2024; Rani et al., 2024; Rasool et al., 2024; Shafqat et al., 2024; Xavier, 2016) has narrowed the focus by us on cities that are classified by the World Bank as low- and middle (lower and upper)-income countries (2016), and with population of inhabitants in Asia, the Caribbean, Africa, and Latin America ranging from 300,000 to over ten million. The Global South or developing countries are those countries that are referred to by this paper.

The Need for Interdisciplinary Collaboration

We emphasize the importance of ecological and engineering objectives are trade-offs between these objectives and are navigated by interdisciplinary collaboration, ultimately enabling us to engineer both *with and for* biodiversity.

Sustainability Training Program and Citizen Science

The HSBC Group initiated a global program Sustainability Training Program (STP), leading to the implementation and designing being responsible with Earthwatch. HSBC staff get awareness of sustainability, which is the aim of this program; sustainable business practices are supported to take action by incentivizing them and sustainable corporate growth is supported and encouraged by them by everyday action. Tianfu and Nansha wetlands carried out 13 measurement events in 2018 and 2019, and 251 HSBC citizen scientists are gathering a total of 2664 monitoring datasets trained by NIGLAS researchers.

The Role of Wetlands as Blue–Green Infrastructure

Ecosystem Services Provided by Wetlands

The most common natural blue–green infrastructure are the wetlands and provide well-defined ecosystem services, including

(a) Flood prevention/mitigation.
(b) Food provision.
(c) Conservation of biodiversity.
(d) Improvement in water quality and microclimate.
(e) Cultural/social functions.

The surface runoff water quality improvement is related to mechanisms that are explored by many studies in wetlands, such as phosphorus removal by soil absorption and nitrate loss by denitrification.

Challenges in Wetland Management

However, development pressures and competing and multiple land uses are contained by many wetlands. Different wetland uses are related to relative benefits on water quality mitigation compared by a few studies. Integrated wetland management policy is created by decision-makers who are allowed by such information. In the present study, two large multiuse wetlands were chosen to compare the effects of wetland utilization compared by choosing two large multiuse wetlands in the present study on its capacity to poor water quality mitigated in incoming waters. Major blue–green infrastructure in areas are Nansha (Zhang et al., 2021) wetland in the Pearl Delta and Tianfu wetland in the Yangtze under high development. Wetland management of different types is possibly examining national and provincial decision-makers who get important and unique opportunity in relation to a key ecosystem service with better tools to manage these important landscapes. Furthermore, robust scientific data is clearly needed in which multiuse wetlands allow us to compare that dominate eastern and southern China.

Discussion

Ecosystem services and functioning are maintained by more species, which is indicated by our results than previous studies suggested that have either

(a) Considered that one function is promoted by only the number of species under one set of environmental conditions or.
(b) Separately considered that ecosystem functioning is provided by the importance of biodiversity across multiple places years environmental changes or functional scenarios.

Therefore, although species may functionally appear redundant when under one set of environmental conditions one function is considered, a changing world has many species that are needed to maintain multiple functions at multiple places and times.

Worldwide declining of biodiversity is rapid, and there is consensus that ecosystem services and functioning also could be decreased. Ecosystem services provision is needed to sustain with few or many species in an ecosystem but still this is remaining though and unclear. It has been hypothesized that most species would promote ecosystem services promoted by mostly species if environmental changes, functions, places, and times were considered 9; however, all of these factors are not considered by the previous study together. Here, we show that of biodiversity experiments, almost 17 have 84% of the 147 grassland plant species studied to promote functioning of ecosystem at least once. Ecosystem functioning is promoted by different species during different places, in different years, for different environmental changes and under different functions (Majeed, Noor, et al., 2024; Muhammad, Basit, Majeed, et al., 2024; Naveed, Zahid, et al., 2024; Rasool et al., 2025). Furthermore, the species needed to provide multiple functions within 1 year were not the same as those needed to provide one function during multiple years.

Engineered and semi-natural elements are combined by blue–green infrastructure (BGI), habitat provision, heat mitigation, water purification, and stormwater management that are multifaceted benefits offered by this BGI. However, ecological potential of engineering goals is overlooking and prioritized by the current design of BGI. BGI designs have integrated both ecological and engineering objectives to enhance biodiversity and performance. Ecological processes, abundance, and importance of species diversity (Majeed et al., 2020; Naveed, Majeed, et al., 2024; Rasool et al., 2024) are emphasized by this interdisciplinary literature review to improve lower management costs and engineering resilience and performance.

Nevertheless, the developed countries, particularly Europe and North America in cities, have been conducted by mostly existing studies of GBI. Developing countries are less known to address the challenges and key environmental issues; as a result of these solutions, the urban GBI is applied on the ground. Innovative solutions and major trends in developing countries are covered by current research; consequently, the academic literature on urban GBI was investigated.

Future Recommendations

Green and blue infrastructure (GBI) optimized by technological interventions should be prioritized for equitable distribution of benefits to drive policy reforms and enhancement of community engagement. Ecological and engineering objectives are integrated by future efforts in the design of GBI, thereby enhanced species diversity is fostering robust ecological processes and resilience. Sustainable funding sources for urban green infrastructure are ensured by exploring innovative financing mechanisms, while inclusive growth in UGI implementation in which equity issues are addressed. Green infrastructure is imperative in which urban biodiversity (Majeed, Noor, et al., 2024; Naveed, Memon, et al., 2024) is increased and protected, and urban biodiversity explored best practices and case studies through which valuable future initiatives will be provided. Ecosystem services integrate biodiversity and spatial conservation prioritization methods in GBI design that are crucial to enhance tools to maximize benefits. Biodiversity-based solutions are emphasized by future projects in urban green infrastructure, and the impacts of species diversity are assessed to inform conditions and design criteria on ecosystem services. Infrastructures are supported by the impact of biodiversity that must be continually evaluated on urban ecosystems, with urban green infrastructure, which is an evolving approach in which biodiversity is focused. Overall outcomes will be strengthened by the integration of biodiversity with other urban sustainability goals. Diverse regions have implemented the proposed methodology that is essential for refining best practices for validation. Additionally, the effectiveness of GBI in biodiversity conservation efforts assessment should include strategies under climate-changing scenarios for protecting species. Resilience will be enhanced for future management practices by selecting different species, while understanding the impacts of public health of urban biodiversity and community well-being is promoted by infrastructure. Health promotion is prioritized in urban planning by preventive measures, and sustainable outcomes have the integration of financial and ecological benefits by rainwater management parameters.

Conclusion

Environmental justice, resilience, and urban sustainability are enhanced by BGIs. BGIs' functionality is optimized by crucial advanced technology in urban environments. Enhancement of BGIs design is done by the integration of ecological and engineering objectives. Ecological and engineering goals are effectively balanced by interdisciplinary collaboration, which is essential. Environmental protection and human development are promoted by integrated renewable energies. Furthermore, preservation of the Villa Marshes ecosystem and enhancement of awareness is done by green corridors. Sustainable urban landscapes, biodiversity-rich, and UGI foster resilient. Effective UGI implementation needs adaptive management. Improvement in human health and well-being and critical ecosystem services is provided by restoring and increasing biodiversity in cities. Enhancement and conservation of urban biodiversity take place by effective tools like multifunctional green infrastructure. Pollution levels are reduced by the preservation of biodiversity. Nature conservation with societal development is harmonized by green infrastructure. This holistic approach in which environmental degradation is not only mitigated but also community has a foster deeper sense among urban residents and connection to nature. Biodiversity conservation is vital in the role of urban areas. Infrastructures are supported by biodiversity significantly to improve citizen relations and urban biodiversity. Effective green infrastructure strategies enhanced the urban biodiversity, more resilient ecosystems can be led by us, and city dwellers are ultimately benefited by both qualities of life and environment. Recreational spaces, temperature regulation, and water and air purification are the essential ecosystem services that promote mental well-being, which could be provided by such strategies. In conclusion, thriving and sustainable urban environments are created by investing in these strategies for future generations.

References

Ali, H., Muhammad, Z., Majeed, M., Aziz, R., Khan, A., Mangrio, W. M., & Al Dughairi, A. A. (2023). Vegetation diversity pattern during spring season in relation to topographic and edaphic variables in subtropical zone. *Botanical Studies, 64*(1), 25. https://doi.org/10.1186/s40529-023-00398-5

Arif, U., Bhatti, K. H., Ajaib, M., Wagay, N. A., Majeed, M., Zeb, J., Hameed, A., & Kiani, J. (2021). Ethnobotanical indigenous knowledge of Tehsil Charhoi, District Kotli, Azad Jammu and Kashmir, Pakistan. *Ethnobotany Research and Applications, 22*, 24. https://doi.org/10.32859/ERA.22.50.1-24

Arshad, F., Waheed, M., Harun, N., Fatima, K., Khan, B. A., Fatima, K., Abbas, Z., Jabeen, S., & Majeed, M. (2022). Indigenous farmer's perception about fodder and foraging species of semi-arid lowlands of Pakistan: A case study of District Kasur, Pakistan. *Taiwania, 67*(4), 510–523. https://doi.org/10.6165/tai.2022.67.510

Ashrafzadeh, M. R., Khosravi, R., Adibi, M. A., Taktehrani, A., Wan, H. Y., & Cushman, S. A. (2020). A multi-scale, multi-species approach for assessing effectiveness of habitat and connectivity conservation for endangered felids. *Biological Conservation, 245*, 108523.

Aziz, A., Anwar, M. M., Majeed, M., Albanai, J. A., Almohamad, H., & Abdo, H. G. (2023). Quantifying the impacts of urbanization on urban green, evidences from Maga City, Lahore Pakistan. *Discover Sustainability, 4*(1), 1–11. https://doi.org/10.1007/s43621-023-00169-z

Aziz, A., Anwar, M. M., Majeed, M., Fatima, S., Mehdi, S. S., Mangrio, W. M., Elbouzidi, A., Abdullah, M., Shaukat, S., Zahid, N., Mahmoud, E. A., Casini, R., Yessoufou, K., & Elansary, H. O. (2023). Quantifying landscape and social amenities as ecosystem services in rapidly changing peri-urban landscape. *Land, 12*(2), 477. https://doi.org/10.3390/land12020477

Baliello, A. (2020). *The role of road pavements in the phenomenon of urban Heat Island*. Università degli Studi di Trieste.

Bashir, S. M., Altaf, M., Hussain, T., Umair, M., Majeed, M., Mangrio, W. M., Khan, A. M., Gulshan, A. B., Hamed, M. H., Ashraf, S., Amjad, M. S., Bussmann, R. W., Abbasi, A. M., Casini, R., Alataway, A., Dewidar, A. Z., Al-Yafrsi, M., Amin, M. H., & Elansary, H. O. (2023). Vernacular taxonomy, cultural and ethnopharmacological applications of avian and mammalian species in the vicinity of Ayubia National Park, Himalayan Region. *Biology, 12*(4), 609. https://doi.org/10.3390/biology12040609

Bhattacharya, R., Bhattacharya, R., Majeed, M., Bhandari, S., Aziz, R., Sinha, D., & Niyogi, S. S. (2023). Environmental rehabilitation of industrial waste dumping site. In *Biohydrometallurgical processes* (pp. 44–89). CRC Press. https://doi.org/10.1201/9781003451457-4

Boley, B. B. (2009). *Geotourism in the crown of the continent: Developing and testing the geotourism survey instrument*. The University of Montana.

Coutts, C., & Hahn, M. (2015). Green infrastructure, ecosystem services, and human health. *International Journal of Environmental Research and Public Health, 12*(8), 9768–9798.

Diamond, J. (1988). Factors controlling species diversity: Overview and synthesis. *Annals of the Missouri Botanical Garden, 75*, 117–129.

Drosou, N., Soetanto, R., Hermawan, F., Chmutina, K., Bosher, L., & Hatmoko, J. U. D. (2019). Key factors influencing wider adoption of blue–green infrastructure in developing cities. *Water, 11*(6), 1234.

Francis, R. A., & Chadwick, M. A. (2013). *Urban ecosystems: Understanding the human environment*. Routledge.

Gamon, J. A., Somers, B., Malenovský, Z., Middleton, E. M., Rascher, U., & Schaepman, M. E. (2019). Assessing vegetation function with imaging spectroscopy. *Surveys in Geophysics, 40*, 489–513.

Gardner, C. M., Laryea, K. B., & Unger, P. W. (1999). *Soil physical constraints to plant growth and crop production* (Vol. 11). Food and Agriculture Organization.

Greco, I., & Bencardino, M. (2014). The paradigm of the modern city: SMART and SENSEable cities for SMART, inclusive and sustainable growth. In *Computational science and its applications–ICCSA 2014: 14th international conference, proceedings, part II 14* (pp. 579–597). Springer International Publishing.

Haq, S. M., Tariq, A., Li, Q., Yaqoob, U., Majeed, M., Hassan, M., Fatima, S., Kumar, M., Bussmann, R. W., Moazzam, M. F. U., & Aslam, M. (2022). Influence of edaphic properties in determining forest community patterns of the zabarwan mountain range in the Kashmir Himalayas. *Forests, 13*(8), 1214. https://doi.org/10.3390/f13081214

Haq, S. M., Yaqoob, U., Majeed, M., Amjad, M. S., Hassan, M., Ahmad, R., & Morales-de la Nuez, A. (2022). Quantitative ethnoveterinary study on plant resource utilization by indigenous communities in high-altitude regions. *Frontiers in Veterinary Science, 9*, 944046.

Hassan, M., Haq, S. M., Ahmad, R., Majeed, M., Sahito, H. A., Shirani, M., Mubeen, I., Aziz, M. A., Pieroni, A., Bussmann, R. W., Alataway, A., Dewidar, A. Z., Al-Yafrsi, M., Elansary, H. O., & Yessoufou, K. (2022). Traditional use of wild and domestic fauna among different ethnic groups in the Western Himalayas—A cross cultural analysis. *Animals, 12*(17), 2276. https://doi.org/10.3390/ani12172276

Hassan, M., Haq, S. M., Majeed, M., Umair, M., Sahito, H. A., Shirani, M., Waheed, M., Aziz, R., Ahmad,

R., Bussmann, R. W., Alataway, A., Dewidar, A. Z., El-Abedin, T. K. Z., Al-Yafrsi, M., Elansary, H. O., & Yessoufou, K. (2022). Traditional food and medicine: Ethno-traditional usage of fish fauna across the valley of Kashmir: A western Himalayan region. *Diversity, 14*(6), 455. https://doi.org/10.3390/d14060455

Heidari, B., Randle, S., Minchillo, D., & Jaber, F. H. (2023). Green stormwater infrastructure: A critical review of the barriers and solutions to widespread implementation. *Wiley Interdisciplinary Reviews: Water, 10*(2), e1625.

Hobbie, S. E., & Grimm, N. B. (2020). Nature-based approaches to managing climate change impacts in cities. *Philosophical Transactions of the Royal Society B, 375*(1794), 20190124.

Hussain, F., Salam, I. U., Memon, Z. N., Abdullah, M., Abbas, G., Akbar, M., & Moda, H. M. (2024). Occurrence of fungal microbial contamination in drinking water of megacity of Karachi (Pakistan) and their physico-chemical control. *Heliyon, 10*(7), e28926. https://doi.org/10.1016/j.heliyon.2024.e28926

Hussain, S., Raza, A., Abdo, H. G., Mubeen, M., Tariq, A., Nasim, W., Majeed, M., Almohamad, H., & Al Dughairi, A. A. (2023). Relation of land surface temperature with different vegetation indices using multi-temporal remote sensing data in Sahiwal region, Pakistan. *Geoscience Letters, 10*(1), 33. https://doi.org/10.1186/s40562-023-00287-6

Jamil, M. D., Waheed, M., Akhtar, S., Bangash, N., Chaudhari, S. K., Majeed, M., Hussain, M., Ali, K., & Jones, D. A. (2022). Invasive plants diversity, ecological status, and distribution pattern in relation to edaphic factors in different habitat types of district Mandi Bahauddin, Punjab, Pakistan. *Sustainability, 14*(20), 13312. https://doi.org/10.3390/su142013312

Khan, A. M., Altaf, M., Hussain, T., Hamed, M. H., Safdar, U., Ayub, A., & Mahmoud, E. A. (2024). Ethnopharmacological uses of fauna among the people of Central Punjab, Pakistan. *Frontiers in Veterinary Science, 11*, 1351693. https://doi.org/10.3389/fvets.2024.1351693

Khan, A. M., Li, Q., Saqib, Z., Khan, N., Habib, T., Khalid, N., Majeed, M., & Tariq, A. (2022). MaxEnt modelling and impact of climate change on habitat suitability variations of economically important chilgoza pine (Pinus gerardiana wall.) in South Asia. *Forests, 13*(5), 715. https://doi.org/10.3390/f13050715

Khoja, A. A., Haq, S. M., Majeed, M., Hassan, M., Waheed, M., Yaqoob, U., Bussmann, R. W., Alataway, A., Dewidar, A. Z., Al-Yafrsi, M., Elansary, H. O., Yessoufou, K., & Zaman, W. (2022). Diversity, ecological and traditional knowledge of Pteridophytes in the Western Himalayas. *Diversity, 14*(8), 628. https://doi.org/10.3390/d14080628

Liu, H. Y., Jay, M., & Chen, X. (2021). The role of nature-based solutions for improving environmental quality, health and well-being. *Sustainability, 13*(19), 10950.

Lu, J. W., Svendsen, E. S., Campbell, L. K., Greenfeld, J., Braden, J., King, K. L., & Falxa-Raymond, N. (2011). Biological, social, and urban design factors affecting young street tree mortality in New York City. *Cities and the Environment, 3*(1), 5.

Majeed, M., Bhatti, K. H., & Amjad, M. S. (2021). Impact of climatic variations on the flowering phenology of plant species in Jhelum district, Punjab, Pakistan. *Applied Ecology and Environmental Research, 19*(5), 3343–3376. https://doi.org/10.15666/aeer/1905_33433376

Majeed, M., Bhatti, K. H., Amjad, M. S., Abbasi, M., Id, R. W. B., Nawaz, F., Rashid, A., Mehmood, A., Id, M. M., Khan, W. M., & Id, S. A. (2020). Ethnoveterinary uses of Poaceae in Punjab, Pakistan. *PLoS One, 15*, e0241705. https://doi.org/10.1371/journal.pone.0241705

Majeed, M., Bhatti, K. H., Pieroni, A., Sõukand, R., Bussmann, R. W., Khan, A. M., Chaudhari, S. K., Aziz, M. A., & Amjad, M. S. (2021). Gathered wild food plants among diverse religious groups in Jhelum District, Punjab, Pakistan. *Foods, 10*(3), 594. https://doi.org/10.3390/foods10030594

Majeed, M., Gulshan, A. B., Muhammad, M., Hussain, M., Aziz, R., Abdullah, M., & Tariq, M. (2024). An investigation into the reactions of biochar in heavy metal polluted soils. In *Biochar-assisted remediation of contaminated soils under changing climate* (pp. 247–276). Elsevier. https://doi.org/10.1016/B978-0-443-21562-9.00010-4

Majeed, M., Khan, A. M., Habib, T., Anwar, M. M., Sahito, H. A., Khan, N., & Ali, K. (2022). Vegetation analysis and environmental indicators of an arid tropical forest ecosystem of Pakistan. *Ecological Indicators, 142*, 109291. https://doi.org/10.1016/j.ecolind.2022.109291

Majeed, M., Lu, L., Anwar, M. M., Tariq, A., Qin, S., El-Hefnawy, M. E., El-Sharnouby, M., Li, Q., & Alasmari, A. (2023). Prediction of flash flood susceptibility using integrating analytic hierarchy process (AHP) and frequency ratio (FR) algorithms. *Frontiers in Environmental Science, 10*, 1037547. https://doi.org/10.3389/fenvs.2022.1037547

Majeed, M., Lu, L., Haq, S. M., Waheed, M., Sahito, H. A., Fatima, S., Aziz, R., Bussmann, R. W., Tariq, A., Ullah, I., & Aslam, M. (2022). Spatiotemporal distribution patterns of climbers along an abiotic gradient in Jhelum District, Punjab, Pakistan. *Forests, 13*(8), 1244. https://doi.org/10.3390/f13081244

Majeed, M., Muhammad, M., Hussain, T., Ali, M., Naz, T., Nawaz, S., & Abbas, G. (2023). Nanopriming-based management of biotic stresses for sustainable agriculture. In *Nanopriming approach to sustainable agriculture* (pp. 263–289). IGI Global. https://doi.org/10.4018/978-1-6684-7232-3.ch012

Majeed, M., Muhammad, M., Nawaz, S., Naz, T., Iqbal, M. M., Zahid, N., & Mehboob, M. (2023). Nanopriming for crop management for sustainable agriculture: An overview. In *Nanopriming approach to sustainable agriculture* (pp. 110–141). IGI Global. https://doi.org/10.4018/978-1-6684-7232-3.ch005

Majeed, M., Noor, W., Rasool, A., Saleem, M., Bashir, I., Fatima, R., & Rafique, A. (2024). Nanotechnology for soil remediation and land restoration. In *Harnessing nanoomics and nanozymes for sustainable agriculture* (pp. 60–77). IGI Global. https://doi.org/10.4018/979-8-3693-1890-4.ch004

Majeed, M., Tariq, A., Anwar, M. M., Khan, A. M., Arshad, F., Mumtaz, F., Farhan, M., Zhang, L., Zafar, A., Aziz, M., Abbasi, S., Rahman, G., Hussain, S., Waheed, M., Fatima, K., & Shaukat, S. (2021). Monitoring of land use–land cover change and potential causal factors of climate change in Jhelum District, Punjab, Pakistan, through GIS and multi-temporal satellite data. *Land, 10*(10), 1026. https://doi.org/10.3390/land10101026

McClean, C. J., Lovett, J. C., Küper, W., Hannah, L., Sommer, J. H., Barthlott, W., & Taplin, J. R. (2005). African plant diversity and climate change. *Annals of the Missouri Botanical Garden, 92*, 139–152.

Mjimba, V. (2019). Green building policy environment. In *The green building evolution* (Vol. 59, p. African Books Collective).

Muhammad, M., Basit, A., Arooj, A., Dixit, G., Majeed, M., Sinha, D., & Li, W. J. (2024). Roles of gibberellins in plant defense against biotic and abiotic stress. In *Plant growth regulators to manage biotic and abiotic stress in agroecosystems* (pp. 183–210). CRC Press.

Muhammad, M., Basit, A., Majeed, M., Shah, A. A., Ullah, I., Mohamed, H. I., & Ghanaim, A. M. (2024). Bacterial pigments and their applications. In *Bacterial secondary metabolites* (pp. 277–298). Elsevier. https://doi.org/10.1016/B978-0-323-95251-4.00019-3

Muhammad, M., Waheed, A., Wahab, A., Majeed, M., Nazim, M., Liu, Y. H., & Li, W. J. (2023). Soil salinity and drought tolerance: An evaluation of plant growth, productivity, microbial diversity, and amelioration strategies. *Plant Stress, 11*, 100319. https://doi.org/10.1016/j.stress.2023.100319

Najam, A. (2020). The view from the south: Developing countries in global environmental politics. In *The global environment: Institutions, law, and policy* (pp. 245–268). CQ Press.

Nasar, S., Muhammad, M., Siddiqi, E. H., Majeed, M., Ameen, N., Ullah, S., & Abdullah, M. (2024). OMICS-based knowledge for achieving food and nutritional security. In *OMICs-based techniques for global food security* (pp. 67–90). Wiley. https://doi.org/10.1002/9781394209156.ch4

Nasar, S., Rasool, A., Ikram, W., Shujat, S., Majeed, M., Muhammad, M., & Saleem, S. (2024). Nano-osmics: A new approach to plant breeding. In *Sustainable agriculture: Nanotechnology and biotechnology for crop production and protection* (p. 129). De Gruyter Brill. https://doi.org/10.1515/9783111234694-008

Naveed, M., Majeed, M., Jabeen, K., Hanif, N., Naveed, R., Saleem, S., & Khan, N. (2024). Plant disease diagnosis with artificial intelligence (AI). In *Microbial data intelligence and computational techniques for sustainable computing* (pp. 217–234). Springer. https://doi.org/10.1007/978-981-99-9621-6_15

Naveed, M., Memon, Z. N., Abdullah, M., Makhdoom, S. I., Azeem, A., Mehmood, S., & Majeed, M. (2024). Use cases and future aspects of intelligent techniques in microbial data analysis. In *Microbial data intelligence and computational techniques for sustainable computing* (pp. 259–280). Springer. https://doi.org/10.1007/978-981-99-9621-6_17

Naveed, M., Zahid, N., Fatima, I., Saleem, A., Majeed, M., Abid, A., & Qasim, A. (2024). Sustainable AI-driven applications for plant care and treatment. In *Microbial data intelligence and computational techniques for sustainable computing* (pp. 235–258). Springer. https://doi.org/10.1007/978-981-99-9621-6_16

Nguyen, T. T. (2022). Ecosystem services assessment of blue-green and grey infrastructure to support planning of sustainable and resilient cities.

Nguyen, T. T., Meurk, C., Benavidez, R., Jackson, B., & Pahlow, M. (2021). The effect of blue-green infrastructure on habitat connectivity and biodiversity: A case study in the Ōtākaro/Avon River catchment in Christchurch, New Zealand. *Sustainability, 13*(12), 6732.

Rani, J., Rasool, A., Sher, H., Fatima, R., Hussain, J., Majeed, M., & Hussain, K. (2024). Precision farming with nanoscale sensors: Monitoring soil, water, and crop health. In *Harnessing nanoomics and nanozymes for sustainable agriculture* (pp. 78–99). IGI Global. https://doi.org/10.4018/979-8-3693-1890-4.ch005

Rasool, A., Feroz, E., Qayyum Khan, Z., Qayyum Khan, M., Majeed, M., & Kousar, S. (2025). Sustainable agriculture: A nanoosmic-based management. In V. Rajput, A. Singh, K. Ghazaryan, T. Minkina, & A. Al-Tawaha (Eds.), *Sustainable agriculture: Nanotechnology, biotechnology, management and food security* (pp. 411–430). De Gruyter. https://doi.org/10.1515/9783111369754-019

Rasool, A., Tariq, M., Asif, M., Shah, G. M., Fatima, R., Majeed, M., & Ozodbek, A. (2024). Nanotechnology for improved crop resilience in challenging environments. In *Harnessing nanoomics and nanozymes for sustainable agriculture* (pp. 149–175). IGI Global. https://doi.org/10.4018/979-8-3693-1890-4.ch008

Roman, L. A., Pearsall, H., Eisenman, T. S., Conway, T. M., Fahey, R. T., Landry, S., & Staudhammer, C. (2018). Human and biophysical legacies shape contemporary urban forests: A literature synthesis. *Urban Forestry & Urban Greening, 31*, 157–168.

Sennett, R., & Sendra, P. (2020). *Designing disorder: Experiments and disruptions in the city*. Verso Books.

Seto, K. C., Sánchez-Rodríguez, R., & Fragkias, M. (2010). The new geography of contemporary urbanization and the environment. *Annual Review of Environment and Resources, 35*(1), 167–194.

Shafqat, A., Rasool, A., Fatima, R., Majeed, M., Rahman, K. U., Narejo, A. H., & Farid, A. (2024). Nanomaterials for sustainable pest and disease management in agriculture. In *Harnessing nanoomics and nanozymes for sustainable agriculture* (pp. 100–125). IGI Global. https://doi.org/10.4018/979-8-3693-1890-4.ch006

Sinha, D., Maurya, A. K., Abdi, G., Majeed, M., Agarwal, R., Mukherjee, R., Ganguly, S., Aziz, R., Bhatia, M., Majgaonkar, A., Seal, S., Das, M., Banerjee, S., Chowdhury, S., Adeyemi, S. B., & Chen, J. T. (2023). Integrated genomic selection for accelerating breeding programs of climate-smart cereals. *Genes, 14*(7), 1484. https://doi.org/10.3390/genes14071484

Tariq, A., Mumtaz, F., Majeed, M., & Zeng, X. (2023). Spatio-temporal assessment of land use land cover based on trajectories and cellular automata Markov modelling and its impact on land surface temperature of Lahore district Pakistan. *Environmental Monitoring and Assessment, 195*(1), 114. https://doi.org/10.1007/s10661-022-10738-w

Tassadduq, S. S., Akhtar, S., Waheed, M., Bangash, N., Nayab, D. E., Majeed, M., Abbasi, S., Muhammad, M., Alataway, A., Dewidar, A. Z., Elansary, H. O., & Yessoufou, K. (2022). Ecological distribution patterns of wild grasses and abiotic factors. *Sustainability, 14*(18), 11117. https://doi.org/10.3390/su141811117

Trodahl, M. I., Jackson, B. M., Deslippe, J. R., & Metherell, A. K. (2017). Investigating trade-offs between water quality and agricultural productivity using the land utilisation and capability indicator (LUCI)–a New Zealand application. *Ecosystem Services, 26*, 388–399.

Ullah, I., Aslam, B., Shah, S. H. I. A., Tariq, A., Qin, S., Majeed, M., & Havenith, H. B. (2022). An integrated approach of machine learning, remote sensing, and GIS data for the landslide susceptibility mapping. *Land, 11*(8), 1265. https://doi.org/10.3390/land11081265

Waheed, M., Arshad, F., Majeed, M., Fatima, S., Mukhtar, N., Aziz, R., Mangrio, W. M., Almohamad, H., Al Dughairi, A. A., Al-Mutiry, M., & Abdo, H. G. (2022). Community structure and distribution pattern of Woody vegetation in response to soil properties in semi-arid Lowland District Kasur Punjab, Pakistan. *Land, 11*(12), 2145. https://doi.org/10.3390/land11122145

Xavier, S. (2016). Learning from below: Theorising global governance through ethnographies and critical reflections from the global south. *Windsor Yearbook of Access to Justice, 33*, 229.

Zhang, Y., Loiselle, S., Zhang, Y., Wang, Q., Sun, X., Hu, M., & Jing, Y. (2021). Comparing wetland ecosystems service provision under different management approaches: Two cases study of Tianfu wetland and Nansha wetland in China. *Sustainability, 13*(16), 8710.

Heavy Metal Pollution and Its Environmental Impacts: A Review of Remediation Potential on Pb/Zn Smelting Sites

Asad Aziz and Muhammad Mushahid Anwar

Abstract

Smelting activities by heavy metals pose a serious impact on the local environment in an industrial area and threaten the local ecology. This chapter summarizes the data on heavy metals in China's industrial regions in the past years. It has been investigated that the heavy metals released from natural and anthropogenic activities negatively impact neighborhoods, habitats, and human health. These approaches depend on and vary from geochemical properties, contaminants, nature of minerals, active or nonactive silicates, and the occurrence of metals. The geochemical properties are mineral liberation analyzer (MLA) and X-ray photoelectron spectroscopy (XPS). The contamination due to the leaching of these heavy metals and chemicals by leaching causes acidification and decreases the pH of the soil. The large-scale survey and small-scale monitoring have proved to be a potential remediation for ecological balance and ecosystem health. The non-ferrous contaminants, for example, Pb and Zn, are major soil pollutants. Therefore, in China, the strategies should primarily be focused on controlling emissions, especially Zn and Pb, along with other pollutants. As a global producer and developer of socio-economic development, China is a heavy consumer of metals and needs to reshape the environmental policies for environmental development.

Keywords

China · Contaminants · Environment · Ecosystem · Ecosystem services · Heavy metals · Pollutant · Smelting · Emission · Pollutants

Introduction

Historical Context

China, as a global leader in the production and consumer of metals and metalloids with 12% of total minerals continuously, improves its socio-economic development (Zhang et al., 2022). This socio-economic and industrial development is done by mineral extraction from the nature and environment (Ghanem & Rukia, 2020). On the other hand, this industrial surge also increases the deposition of heavy metals/metalloids that pose serious environmental threats to local habitat, environment, ecosystem, and biodiversity (Okenmuo & Ewemoje, 2023; Shen et al., 2017).

A. Aziz (✉) · M. M. Anwar
Department of Geography, University of Gujrat, Gujrat, Punjab, Pakistan
e-mail: asadaziz@uog.edu.pk; mushahid.anwar@uog.edu.pk

M. M. Anwar, A. Aziz (eds.), *Sustainable Landscape for Urban Resilience*, Sustainable Development Goals Series, https://doi.org/10.1007/978-3-031-99836-2_11

Smelting activities are a major concern for the environment (Žibret et al., 2018), groundwater, and its quality, ecosystem, biodiversity, and habitat in affected areas. According to a study, southwestern China is the most abundant area historically regarding dispersion, distribution, and mobility of Zn due to industrial activities (Lin et al., 2015). These major targeted sources of Zn and Pb have harmful impacts on local ecological patterns of life (Angon et al., 2024) and the environment due to fume deposition smelting and smelting slags. According to their distribution pattern, the Pb is largely retained in smelting slags; meanwhile, the Zn is predominantly recuperated (Hu, Jin, & Kavan, 2014). This slag poses a long-term effect on biodiversity, ecosystem, environment, and groundwater table compared to contaminated soils because metals in contaminated soil are more mobile compared to slag (Moghimi Dehkordi et al., 2024).

The geographical distribution of Zn and Pb in the region varies in scale depending on the geology, anthropogenic activities, and industry. Figure 1 provides the geographical distribution of Zn and Pb in the country.

In vertical transportation, the Zn is more dominant than Pb (Che et al., 2023). The nonferrous smelting sites cause heavy soil pollution with varying degrees of soil contamination and environmental degradation (Qi et al., 2023). These heavy metals alter the local soil profile and soil fertility in an affected area. In an affected area, these heavy metals cause serious ecological problems, which lead to negative impacts on human health. The ecological issues may involve the migration and vertical movement through leaching from surface deep layers of soil reaching up to 5 m in depth with contamination (Adnan et al., 2024). By combining with deep water or water table, they may be involved in the food chain and damage the ecological chain of food and environment (Nyiramigisha et al., 2021).

Literature Review

Geochemical and Industrial Drivers

The mine tailings are a leading and large driver of lead/zinc (Pb/Zn) with improper deposition at a large scale and challenge the local environment (Li et al., 2023; Sun et al., 2022). The spatial distributions of these metals or deposition are heterogeneous in the main mainland of China. In the northwestern Yunnan Province, Lanping Basin is a massive sediment-hosted Jinding zinc-lead deposit (Pizarro Barraza et al., 2024). However, the historical debate on the genesis of these deposits and heavy metals is still controversial. The historical background of these deposits (Zn, Pb, Sr, Ag, Cd, TI) dating back to the late Paleozoic to Mesozoic in southwestern China of the Sanjiang Metallogenic Domain (SMD) is multistage opening and closing of Tethys oceans (Tang et al., 2017; Wuana & Okieimen, 2011). Similarly, in southwestern China, the Emeishan large igneous province (ELIP), Yangtze Block is a big deposit with >400 carbonate-hosted epigenetic Pb–Zn according to research by Chen et al. (2022), Kong et al. (2017), and He et al. (2022). The giant Upper Yangtze is also another deposition site of Zn and Pb with >20 Mt. base metals (Nikolova et al., 2023; Rouhani et al., 2024). In the case of the southeastern part, Banbanqiao, Shaojiwan, Maozhachang, Tianqiao, Nayongzhi, Maozhachang, Qingshan, Yunluhe, Guanziyao, and Yinchangpo are the major depositors of these metals (Zhou et al., 2018).

Geochemical Factors of Pb and Zn

The project named China Geochemical Baseline (CGB) (2008–2012) provides data on the continental scale of lead (Pb) concentrations in the pedosphere (Cacciuttolo et al., 2023; Yao et al., 2014). According to this project, the concentration of Pb is predominantly high in southeast-

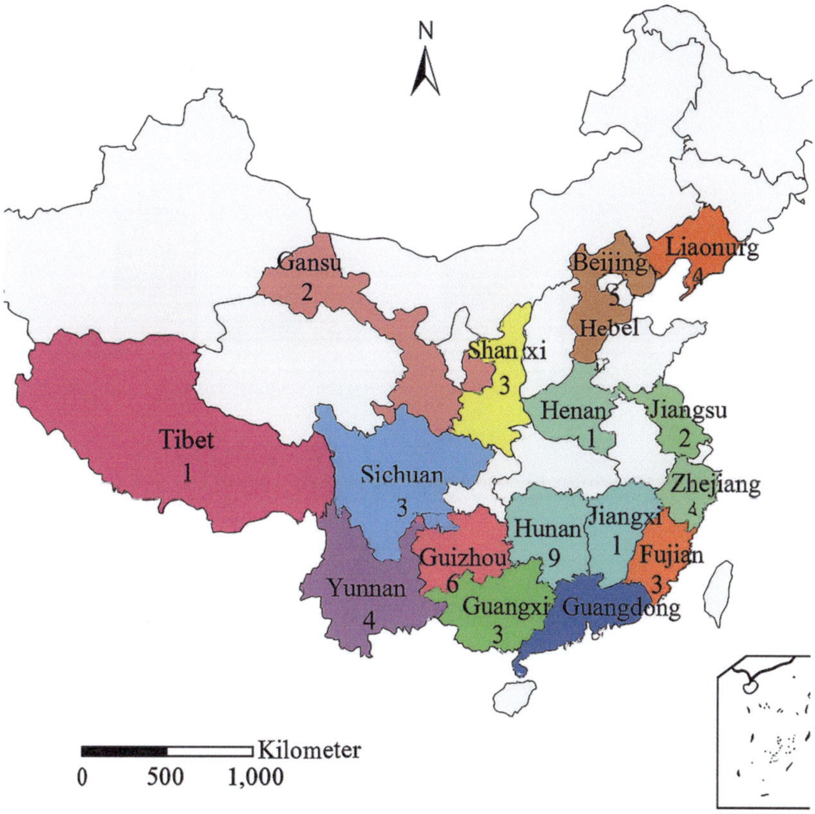

Fig. 1 The geographical distribution of Zn and Pb in China. (Source: Lei et al., 2016)

ern China and comparatively low in northwestern China. The major geochemical factors include the landscape, presence of mineralization, climate, parent rocks, and their geology (Liu, Li, et al., 2020; Shu et al., 2005). The high concentration of Pb is most linked with climate and geology according to Ashraf et al., 2021. While arid and semi-arid regions in northern and northwestern China have low concentrations. According to geochemical characteristics throughout China, the median Pb concentration of Pb is 22.1 mg/kg (Chaudhari et al., 2024). This concentration varies from 15.6 and 30 mg/kg in the main region of the country (Wang, Han, et al., 2019). Geochemically, the permissible limits of Pb and Zn in soil are the highest among other heavy metals followed by plants (Jadoon & Malik, 2019). Figure 2 provides a comparison of lead and zinc among the heavy metals.

Zn or industrial Zn is an important element of organisms and for the industry. On the other hand, it is also an environmental pollutant in China. To date, the study on Zn accumulation on the Chinese agricultural scale is still not well defined. On an industrial scale, the concentration of Zn in China is relatively high (86.14 $mg \cdot kg^{-1}$) compared to the world (Rashid et al., 2023). The total inventory was calculated at 3.08×10^{10} tons in the country, mainly in Yunnan province due to different industrial activities such as smelting, mining, sewage, and irrigation. However, this concentration varied in different regions due to changes in landscape, climate, soil type, and agricultural activities. Zn as a chemical pollutant has adverse impacts on human health as a health hazard ranging from hazard index (HI), > adult males (1.23), > adult females (1.39), and children (2.40). Even the high value of HI is found to be 26.77 in children (Xingyuan Li et al., 2021).

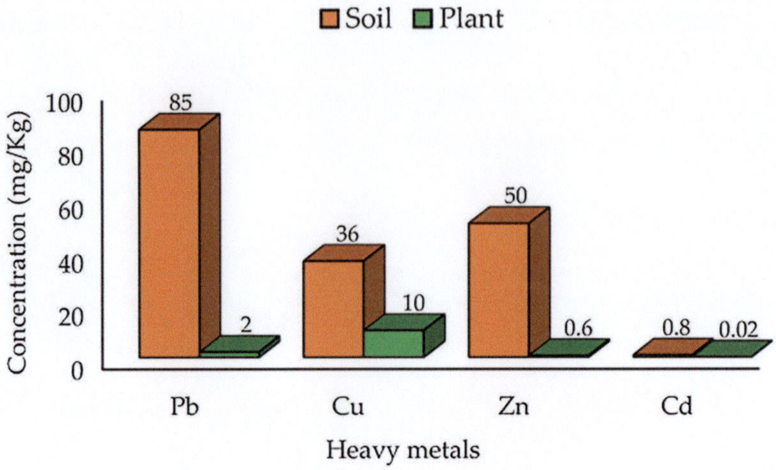

Fig. 2 Heavy metal concentration in soil and plants. (Source: Noulas et al., 2018)

Industrial Practices

Anthropogenic activities, industry, and agricultural practices are the major drivers of high concentrations of heavy metals, including Pb and Zn (Peng & Yu, 2024). The industrial development of the country continuously increases the concentration of these metals in the soil and atmosphere day by day. The Feng County is located southwest of Baoji City and Shaanxi Province of China with four river reserves of 4.5 million tons enriched in Pb and Zn (Liu et al., 2021). Over the past 30 years, Feng County is mainly affected by Zn due to large-scale emission of pollutants by the industry (Bai et al., 2023). In 2012, the acute environmental pollution accidents caused by heavy metals (HM) occurred in the country (Nriagu, 2019). In the country, the spatio-temporal distribution and potential of eco-toxicological risks are still unknown due to limited research on these metals, which is why the soil's physicochemical parameters are still unknown (Wang et al., 2015).

Ecological Risks

With adversative consequences on the human body, Pb and Zn are highly toxic metals for the environment, ecosystem, human health and body, nervous system, and bones (Briffa et al., 2020). The major source of these toxic heavy metal contaminants in roadside soils is vehicular exhaust, which eventually affects agriculture and vegetation as soil contamination (Negahban & Mokarram, 2021). In Guangxi province of China at Xingping Lead–Zinc Mine, a tailing dam in 1976 led to heavy metal concentrations of Pb and Zn into downstream agricultural lands in Side Village as shown in Fig. 3 (Liu et al., 2023). The results of soil pollution assessments showed that the present concentration presented high ecological risk grades (Liu, Wang, et al., 2020). The soil contamination by Zn causes chlorosis of younger leaves and reduces root growth. Because of extreme neurotoxicity, carcinogenicity, and genotoxicity, the remediation of the smelting sites is urgently required (Tchounwou et al., 2012). For underground water, the local environment and agricultural soil the non-ferrous smelting processes and mining are the major releases of toxic metal(loid)s (TMs). Publicly available data in China revealed that approximately 455 contaminated site remediation projects have been initiated across the country (Li et al., 2017). At total of 188 projects out of these 455 are associated with the remediations of heavy metals (Padhye et al., 2023). The impacts of these heavy metals are diverse from hydrological conditions to the soil, ecosystem, and air quality.

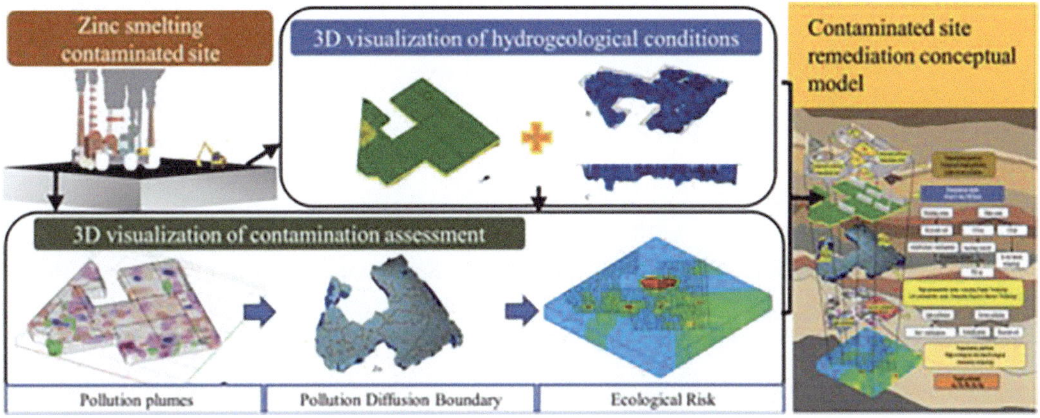

Fig. 3 3D visualization of zinc smelting contamination assessment. (Source: Zeng et al., 2022)

Due to heavy investment and major environmental concerns for planning and development, contaminated site remediation is getting more and more important among policymakers and scientists and is a hot topic for research interest (Awewomom et al., 2024; Geng et al., 2024). This remediation is diverse and multidimensional, and 3D visualization provides a forceful tool for contamination site assessments. The 3D visualization models depicted the highly contaminated industrial sites according to the research by Zhang et al. (2011) and provided visualized diffusion mechanisms to determine the affected boundary and further direction for pre- and post-disaster management. For example, Fig. 1 provides an overview of the contaminated site of Zn for its assessment and remediation from hydrological, environmental, and ecological perspectives (Zeng et al., 2022).

Water Pollution

Mining and smelting are important sources of economic growth for an industrial country. As a leading industrial country in China, there is a large usage of minerals, including Cd, Pb, and Zn for commercial demands (Adnan et al., 2022). The major concentrated areas in China include northwestern Henan province, Guizhou and Sichuan provinces, Yunnan, northwestern Henan province, the region of Liaoning province near Bohai, northern Guangdong, central Guangxi province, and Sichuan and Gansu provinces (Zhang et al. 2011). These heavy metals affect the water, groundwater table, and environment through slag percolation, atmospheric deposition, and sewage irrigation and threaten the cultivated land around a smelter or mining site. Due to the high demand for Pb and Zn, a large amount of waste in gaseous and polluted water is generated and discharged into the open environment, ultimately causing toxicity in water and soil (Zang et al., 2020). The freshwater lakes located in the central and eastern sides of the country are at extremely high ecological risks to heavy metals and metalloids (Zhang et al., 2023). During the past two decades due to industrial development pollution there is increase in spikes air pollution, including zinc (Zn), chromium (Cr), copper (Cu), lead (Pb), mercury (Hg), cadmium (Cd), and nickel (Ni). The concentrations of heavy metals in urban lakes are comparatively higher than the rural area lakes (Xia et al., 2020). The industrial activities, domestic combined with mixed discharge, cause the contamination of heavy metals in lakes. Figure 4 provides the geographical distribution of Chinese lakes contaminated with heavy metals.

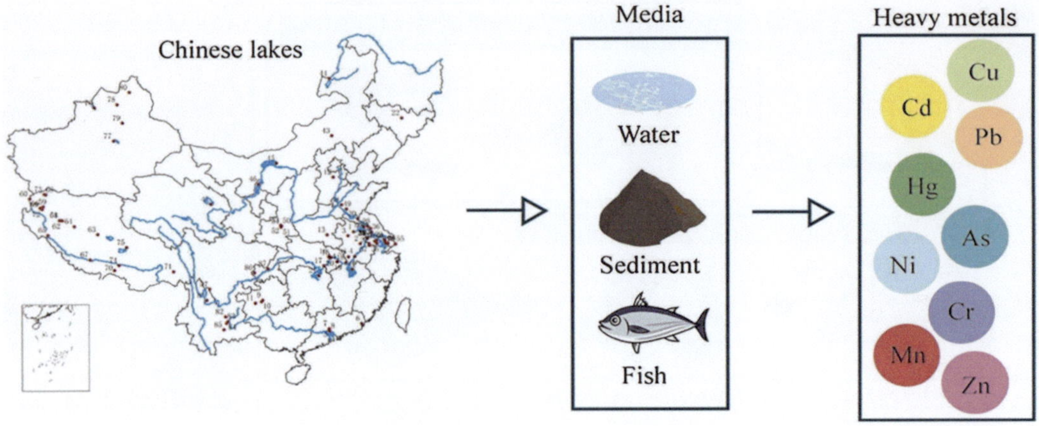

Fig. 4 Geographical distribution of heavy metals-contaminated lakes in China. (Source: Qin & Tao, 2022)

In the freshwater ecosystem, the metal contamination by Zn and Pb is a great threat to insect larvae. Several pieces of evidence show that these threats occurred during the metamorphosis to adults and other aquatic insect of freshwater ecosystem (Lidman et al., 2020). Because of their chemical nature, the Zn and Pb are naturally persistent according to Hussain et al. (2022) and remain under the water as sediments and do not degrade through microbial activity (Mitra et al., 2022). These sediments become polluted as long as any remediation is action (Kapusta & Sobczyk, 2015).

Air Quality

Air quality index (AQI) and air quality health index (AQHI) are associated and generate morbidity in the form of air pollution by different toxic gases, including SO_2 and particulate matter (PM) (Adebayo-Ojo et al., 2023; Shifaw, 2018). The term air quality health index (AQHI) originated in Canada (Chen et al., 2013). In 2023, the air quality index (PM2.5 concentration) of China in Hotan was 6.5 times high compared to the World Health Organization (WHO) standards (Guo et al., 2024). The Hotan Region of Xinjiang is the most polluted and unhealthy city in the country. Throughout the country, the concentration of airborne pollutants can be divided into three periods, including pre-COVID, active COVID, and COVID time, and their chief sources are industry, transport, and energy sectors that were on hold during these periods (Rita et al., 2021). The previous research shows significant variation in the concentration of these pollutants. For example, the concentration of PM2.5 was decreased by 18% during the COVID-19 period (Ghahremanloo et al., 2022; Zheng et al., 2020). Similarly, the concentration of PM10 was decreased by 19% during pre- to active COVID-19 and increased by 23% in post-COVID-19 (Bhatti et al., 2022).

In China, the geological structure of heavy metals is comparatively low according to Hu et al. (2014), but the rapid economic and social activities due to the high population affect the air, air quality, water profile, and aquaculture (Cheng, 2003). Heavy metals such as lead (Pb) and zinc (Zn) have a density greater than 5.0 g/cm^3. Due to their environmental behavior and chemical nature, both often fall in the category of metalloids (Yu et al., 2017). The hot spot areas (Pearl River Estuary (PRE), Hong Kong) in the northern part are comparatively high to southern China (Hu, Jin, & Kavan, 2014). In sum, in urban areas, the heavy metals in contaminated air are a major environmental issue while heavy metals in soil and water are a major issue in rural areas in the country (Xia et al., 2020

Biodiversity Loss

The biodiversity loss by heavy metals is one of the most significant ecological issues in industrial regions and needs attention for a sustainable environment (Briffa et al., 2020; Huang et al., 2019). The heavy metals are accumulated by the plants from atmospheric aerosols by absorption by leaves. Excessive amounts of heavy metals such as Zn and Pb affect the quantity and quality of functioning biota in plants (Laptiev et al., 2024). The pollution index examination shows that Huludao and Zhuzhou are at extreme polluted levels (Xi et al., 2008). The Pb index has been transferred from the Middle East to the West, the provinces Yunnan, Henan, and Hunan Provinces with high degrees (Yu et al., 2020; Zang et al., 2020). Biodiversity and biodegradation are closely associated. The heavy pollutants from industry, transportation, agriculture, and mining pass through different cycles and create an environmental nexus with different phases to form a complicated morphology that enters into the environment and causes degradation (Javan et al., 2024). Figure 5 provides the association of this nexus.

Discussion

Remedial Strategies

The principles and applicability for remediation of soil heavy metals contaminants are classified into two major divisions (Mohamed et al., 2024). This classification includes in situ and ex situ remediation techniques (Liao et al., 2022; Paul et al., 2021). This includes rectification by physical, chemical, thermal, and biological methods and approaches. The most common examples of these techniques are stabilization, electrokinetic extraction, solidification, surface capping, landfilling, encapsulation, soil flushing, surface capping, bioremediation and solidification, and

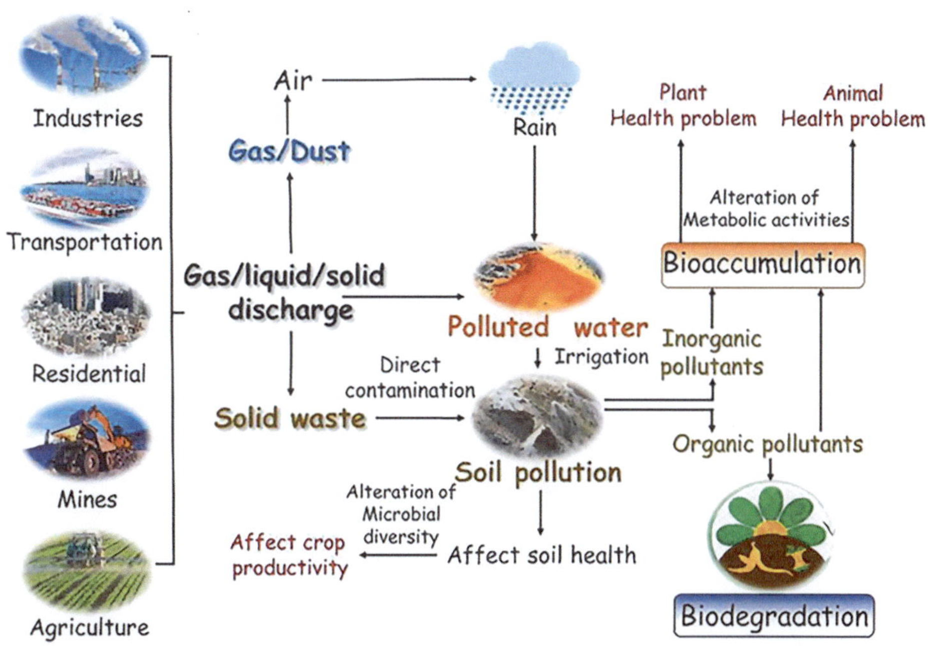

Fig. 5 Heavy metals nexus and environmental association. (Source: Riyam et al., 2022)

vitrification (Gusiatin et al., 2020). These approaches used extraction/removal, containment, and immobilization mechanisms and reduce the contamination effects. With specific advantages and disadvantages in applicability, the techniques are favorable and cost-effective in the developmental stage, duration, field scales, efficiency, and feasibility (Liu et al., 2018; Mosa et al., 2021). According to research, the remediation of the combination of heavy metals Zn and Pb is economical compared to individual metals (Zheng et al., 2024). Figure 6 further elaborates on the remediation process for Zn and Pb.

Integrated Remediation Strategies

The integrated remediation strategy is a multistory concept involving different approaches with multiple techniques depending on soil type, contaminant levels, prioritizing methods, site conditions, required remediation goals, and time and cost for budget constraints (Alsafran et al., 2023; Priya et al., 2023). Integrated remediation strategies combining multiple approaches like phytoremediation and bioremediation as a biological strategy (Asad et al., 2019), solidification, soil washing, electrokinetic remediation, and stabilization for physical and chemical remediation methods (Liu et al., 2024; Nie et al., 2024). Figure 7 elaborates on the future prospective techniques for physical, mechanical, and biological strategies for soil remediation.

Heavy metal pollution is a serious issue for the ecosystem and environment globally (Das et al., 2023). Therefore, this is a burning issue and getting special attention among environmental scientists, ecosystem researchers, policymakers, and scientists for global peace and sustainability.

The selection of the best-used methods depends on multiple procedures to be used for remediation. For example, in bioremediation processes, microbiological metabolism can be used to transfer harmful substances (Azubuike et al., 2016). Similarly, in the case of physical separation processes or methods, specific physical qualities such as density, size, and hydrophobic and magnetic surface features are used to transfer the contaminated media (Poloko, 2019; Stewart, 2020). Meanwhile, in the case of chemical methods, different chemical properties can be used to minimize their impacts. These techniques can be on-site or off-site varying from in situ to ex situ remediation methods (Khatun et al., 2024). Figure 8 provides the classification of these remediation methods. However, the integrated remediation strategies count the natural existence (Kumar et al., 2021), anthropomorphic sources, the physi-

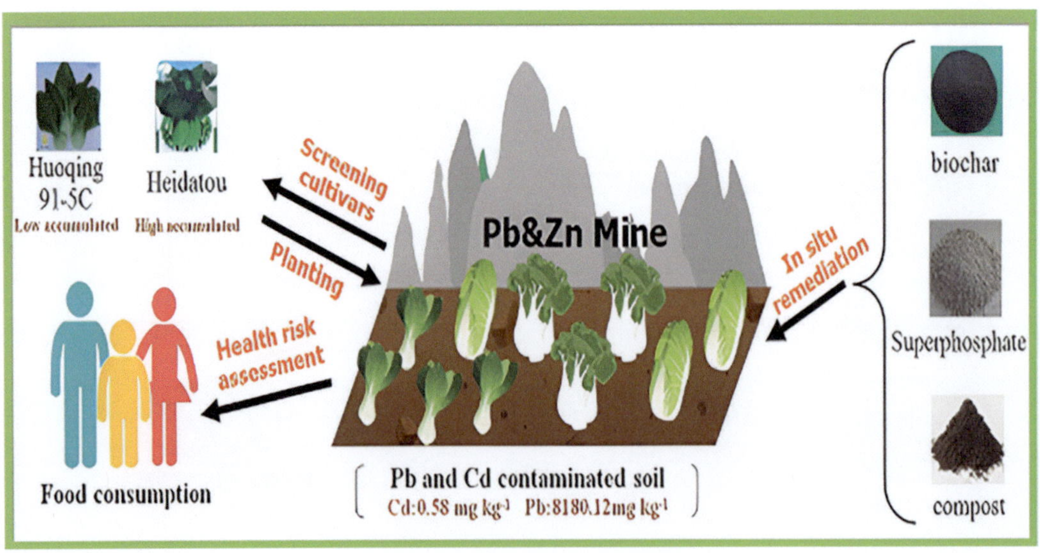

Fig. 6 The remedial strategy of Zn and Pb. (Source: Wang, Yang, et al., 2019)

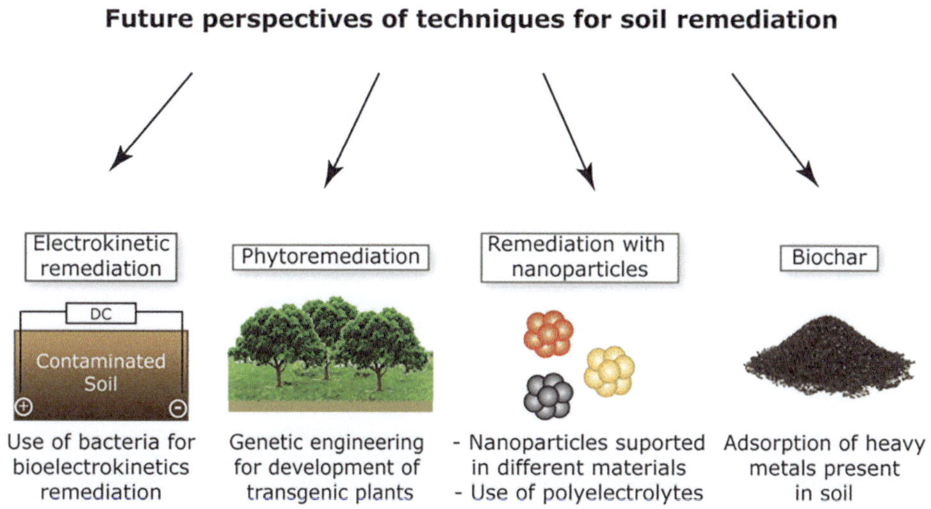

Fig. 7 Future perspectives of biological, chemical, and physical methods for soil contamination. (Source: Souza et al., 2020)

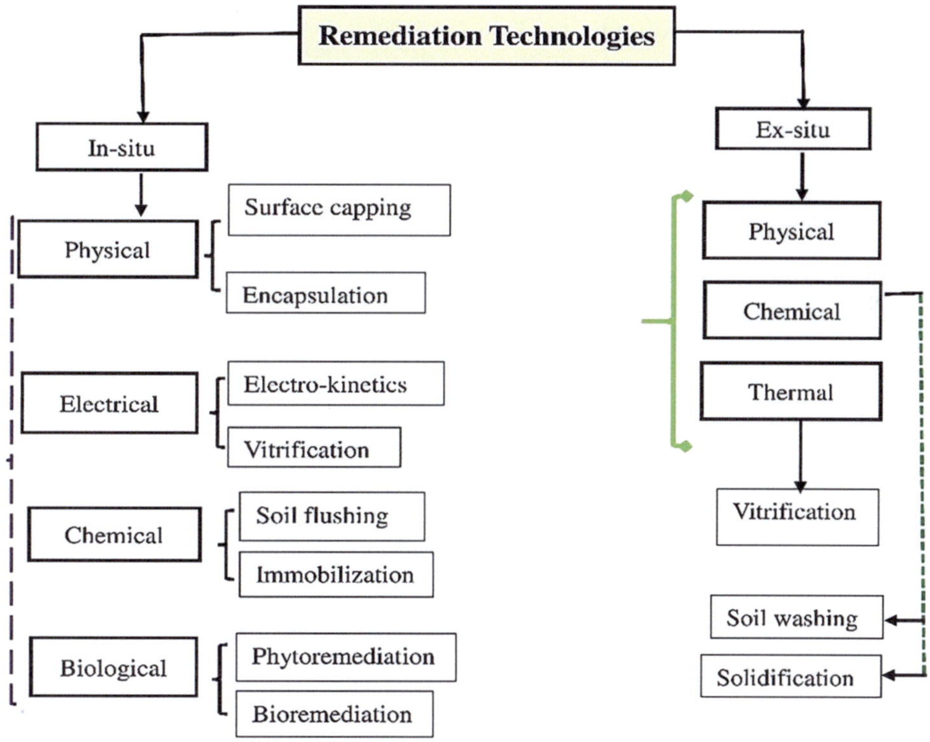

Fig. 8 Classification of remediation methods of heavy metals. (Source: Thakur et al., 2024)

cal environment, and the concentration of heavy metals in the soil or smelting sites (Kuppan et al., 2024; Selvi et al., 2019). Still, there are research gaps that exist in previous research due to varying land morphology, soil type, continent level, and area geographical characteristics.

Conclusions

Smelting for heavy metals poses an influential threat to the environment, human health, groundwater, air, and water at a large geographical scale. The word selecting may be used for less efficient material obtained from historical processing from the industry but usually contains valuable metals. The human activities due to industrial activities demand and increase the usage of metal, and heavy metals gradually in this current era. To ensure environmental health and quality for effective control measures, it is a fundamental goal for scientists and policymakers to consider the remedial potential of biological, chemical, physical, and mechanical techniques and methods.

Depending on the site, conditions, cost, and time, and concentration of heavy metals such as Pb and Zn, these approaches are but not limited to stabilization, electrokinetic extraction, solidification, surface capping, landfilling, encapsulation, soil flushing, surface capping, bioremediation and solidification, and vitrification. The study provides evidence that Pb contamination is associated with non-carcinogenic health risks in children. On the other hand, the high concentration of Zn overwhelms homeostasis, resulting in adverse effects on marine and terrestrial life.

As heavy metals, both Pb and Zn have adverse impacts on human health, environment, water, and local ecology of the region in long-term effects. China, the leading country in the world, needs to restructure the policies for land-use planning and concentrating methods to obtain the valuable materials from the smelting and smelting sites to get useful components for industrial development at international levels. This material can be used by other industries as raw material to generate more profit.

References

Adebayo-Ojo, T. C., Wichmann, J., Arowosegbe, O. O., Probst-Hensch, N., Schindler, C., & Künzli, N. (2023). A new global air quality health index based on the WHO air quality guideline values with application in Cape Town. *International Journal of Public Health, 68*, 1606349.

Adnan, M., Xiao, B., Ubaid Ali, M., Peiwen, X., Zhao, P., Wang, H., & Bibi, S. (2024). Heavy metals pollution from smelting activities: A threat to soil and groundwater. *Ecotoxicology and Environmental Safety, 274*, 116189. https://doi.org/10.1016/j.ecoenv.2024.116189

Adnan, M. A.-O., Xiao, B., Xiao, P., Zhao, P., Li, R., & Bibi, S. (2022). Research progress on heavy metals pollution in the soil of smelting sites in China. *Toxics*. https://doi.org/10.3390/toxics10050231

Alsafran, M., Saleem, M. H., Al Jabri, H., Rizwan, M., & Usman, K. (2023). Principles and applicability of integrated remediation strategies for heavy metal removal/recovery from contaminated environments. *Journal of Plant Growth Regulation, 42*(6), 3419–3440. https://doi.org/10.1007/s00344-022-10803-1

Angon, P. B., Islam, M. S., Kc, S., Das, A., Anjum, N., Poudel, A., & Suchi, S. A. (2024). Sources, effects and present perspectives of heavy metals contamination: Soil, plants and human food chain. *Heliyon, 10*(7), e28357. https://doi.org/10.1016/j.heliyon.2024.e28357

Asad, S. A., Farooq, M., Afzal, A., & West, H. (2019). Integrated phytobial heavy metal remediation strategies for a sustainable clean environment - A review. *Chemosphere, 217*, 925–941. https://doi.org/10.1016/j.chemosphere.2018.11.021

Ashraf, I., Ahmad, F., Sharif, A., Altaf, A. R., & Teng, H. (2021). Heavy metals assessment in water, soil, vegetables and their associated health risks via consumption of vegetables, district Kasur, Pakistan. *SN Applied Sciences, 3*(5), 552. https://doi.org/10.1007/s42452-021-04547-y

Awewomom, J., Dzeble, F., Takyi, Y. D., Ashie, W. B., Ettey, E. N. Y. O., Afua, P. E., & Akoto, O. (2024). Addressing global environmental pollution using environmental control techniques: A focus on environmental policy and preventive environmental management. *Discover Environment, 2*(1), 8. https://doi.org/10.1007/s44274-024-00033-5

Azubuike, C. C., Chikere, C. B., & Okpokwasili, G. C. (2016). Bioremediation techniques-classification based on site of application: Principles, advantages, limitations and prospects. *World Journal of Microbiology & Biotechnology, 32*, 180.

Bai, Y., Zhang, Y., Liu, X., & Wang, Y. (2023). The spatial distribution and source apportionment of heavy metals in soil of Shizuishan, China. *Environmental Earth Sciences, 82*(21), 494. https://doi.org/10.1007/s12665-023-11192-8

Bhatti, U. A., Zeeshan, Z., Nizamani, M. M., Bazai, S., Yu, Z., & Yuan, L. (2022). Assessing the change of ambient air quality patterns in Jiangsu Province of China pre-to post-COVID-19. *Chemosphere, 288*, 132569. https://doi.org/10.1016/j.chemosphere.2021.132569

Briffa, J., Sinagra, E., & Blundell, R. (2020). Heavy metal pollution in the environment and their toxicological effects on humans. *Heliyon, 6*(9), e04691. https://doi.org/10.1016/j.heliyon.2020.e04691

Cacciuttolo, C. A.-O., Cano, D. A.-O., & Custodio, M. A.-O. X. (2023). Socio-environmental risks linked with mine tailings chemical composition: Promoting responsible and safe mine tailings management considering copper and gold mining experiences from Chile and Peru. *Toxics, 11*(5), 462. https://doi.org/10.3390/toxics11050462

Chaudhari, U., Kumari, D., Tyagi, T., Mittal, S., & Sahoo, P. K. (2024). Geochemical signature and risk assessment of potential toxic elements in intensively cultivated soils of south-West Punjab, India. *Minerals, 14*(6), 576. https://doi.org/10.3390/min14060576

Che, B., Shao, C., Lu, Z., Qian, B., & Chen, S. (2023). Mineral requirements for China's energy transition to 2060—Focus on electricity and transportation. *Sustainability, 15*(1), 10585. https://doi.org/10.3390/su15010585

Chen, R., Wang, X., Meng, X., Hua, J., Zhou, Z., Chen, B., & Kan, H. (2013). Communicating air pollution-related health risks to the public: An application of the air quality health index in Shanghai, China. *Environment International, 51*, 168–173. https://doi.org/10.1016/j.envint.2012.11.008

Chen, T., Wen, X.-C., Zhang, L.-J., Tu, S.-C., Zhang, J.-H., Sun, R.-N., & Yan, B. (2022). The geochemical and mineralogical controls on the release characteristics of potentially toxic elements from lead/zinc (Pb/Zn) mine tailings. *Environmental Pollution, 315*, 120328. https://doi.org/10.1016/j.envpol.2022.120328

Cheng, S. (2003). *Heavy metal pollution in China: Origin, pattern and control*. U.S. Department of Health and Human Services.

Das, S., Sultana, K. W., Ndhlala, A. R., Mondal, M., & Chandra, I. (2023). Heavy metal pollution in the environment and its impact on health: Exploring green technology for remediation. *Environmental Health Insights, 17*, 11786302231201259. https://doi.org/10.1177/11786302231201259

Geng, J., Fang, W., Liu, M., Yang, J., Ma, Z., & Bi, J. (2024). Advances and future directions of environmental risk research: A bibliometric review. *Science of the Total Environment, 954*, 176246. https://doi.org/10.1016/j.scitotenv.2024.176246

Ghahremanloo, M., Lops, Y., Choi, Y., Jung, J., Mousavinezhad, S., & Hammond, D. (2022). A comprehensive study of the COVID-19 impact on PM(2.5) levels over the contiguous United States: A deep learning approach. *Atmospheric Environment, 272*, 118944.

Ghanem, S., & Rukia, A. (2020). Distribution of soil organic matter in the coastal region of Syria: A case study. *DYSONA - Applied Science, 1*(2), 57–63. https://doi.org/10.30493/das.2020.234950

Guo, Q., He, Z., & Wang, Z. (2024). The characteristics of air quality changes in Hohhot City in China and their relationship with meteorological and socio-economic factors. *Aerosol and Air Quality Research, 24*(5), 230274. https://doi.org/10.4209/aaqr.230274

Gusiatin, M., Kulikowska, D., & Klik, B. (2020). New-generation washing agents in remediation of metal-polluted soils and methods for washing effluent treatment: A review. *International Journal of Environmental Research and Public Health, 17*, 6220. https://doi.org/10.3390/ijerph17176220

He, Z., Li, B., Wang, X., Xiao, X., Wan, X., & Wei, Q. (2022). The origin of carbonate components in carbonate hosted Pb-Zn deposit in the Sichuan-Yunnan-Guizhou Pb-Zn Metallogenic Province and Southwest China: Take Lekai Pb-Zn deposit as an example. *Minerals, 12*(12), 1615. https://doi.org/10.3390/min12121615

Hu, H., Jin, Q., & Kavan, P. (2014). A study of heavy metal pollution in China: Current status, pollution-control policies and countermeasures. *Sustainability, 6*(9), 5820–5838. https://doi.org/10.3390/su6095820

Hu, H.-P., Deng, Q., Li, C., Xie, Y., Dong, Z., & Zhang, W. (2014). The recovery of Zn and Pb and the manufacture of lightweight bricks from zinc smelting slag and clay. *Journal of Hazardous Materials, 271*, 220–227. https://doi.org/10.1016/j.jhazmat.2014.01.035

Huang, X., Fan, Z., Yan, W. D., Chen, X., Wang, G. J., & Wang, R. J. (2019). Effects of Pb and Zn toxicity on chlorophyll fluorescence and biomass production of Koelreuteria paniculata and Zelkova schneideriana young plants. *Photosynthetica, 57*, 50. https://doi.org/10.32615/ps.2019.050

Hussain, S., Khan, M., Sheikh, T. M. M., Mumtaz, M. Z., Chohan, T. A., Shamim, S., & Liu, Y. (2022). Zinc essentiality, toxicity, and its bacterial bioremediation: A comprehensive insight. *Frontiers in Microbiology, 13*, 900740.

Jadoon, W. A., & Malik, R. N. (2019). Geochemical approach for heavy metals in suburban agricultural soils of Sialkot, Pakistan. *SN Applied Sciences, 1*(2), 161. https://doi.org/10.1007/s42452-019-0167-3

Javan, K., Altaee, A., BaniHashemi, S., Darestani, M., Zhou, J., & Pignatta, G. (2024). A review of interconnected challenges in the water–energy–food nexus: Urban pollution perspective towards sustainable development. *Science of the Total Environment, 912*, 169319. https://doi.org/10.1016/j.scitotenv.2023.169319

Kapusta, P., & Sobczyk, Ł. (2015). Effects of heavy metal pollution from mining and smelting on enchytraeid communities under different land management and soil conditions. *Science of the Total Environment, 536*, 517–526. https://doi.org/10.1016/j.scitotenv.2015.07.086

Khatun, M., Kobir, M., Miah, A., Sarkar, A., & Alam, M. A. (2024). Technologies for remediation of heavy metals in environment and ecosystem: A critical overview of comparison study. *Asian Journal of Environment & Ecology, 23*, 61–80. https://doi.org/10.9734/AJEE/2024/v23i4540

Kong, Z., Wu, Y., Liang, T., Zhang, F., Meng, X., & Lu, L. (2017). Sources of ore-forming material for Pb-Zn deposits in the Sichuan-Yunnan-Guizhou triangle area: Multiple constraints from C-H-O-S-Pb-Sr isotopic compositions. *Geological Journal, 53*, 3019. https://doi.org/10.1002/gj.3019

Kumar, M., Seth, A., Singh, A. K., Rajput, M. S., & Sikandar, M. (2021). Remediation strategies for

heavy metals contaminated ecosystem: A review. *Environmental and Sustainability Indicators, 12*, 100155. https://doi.org/10.1016/j.indic.2021.100155
Kuppan, N., Padman, M., Mahadeva, M., Srinivasan, S., & Devarajan, R. (2024). A comprehensive review of sustainable bioremediation techniques: Eco friendly solutions for waste and pollution management. *Waste Management Bulletin, 2*(3), 154–171. https://doi.org/10.1016/j.wmb.2024.07.005
Laptiev, V., Apori, S. O., Giltrap, M., Tian, F., & Ryzhenko, N. (2024). Bioaccumulation of Cr, Zn, Pb and cu in Ambrosia artemisiifolia L. and Erigeron canadensis L. *Resources, 13*(3), 43. https://doi.org/10.3390/resources13030043
Lei, K., Giubilato, E., Critto, A., Pan, H., & Lin, C. (2016). Contamination and human health risk of lead in soils around lead/zinc smelting areas in China. *Environmental Science and Pollution Research, 23*, 13128–13136. https://doi.org/10.1007/s11356-016-6473-z
Li, R., Yin, Z., & Lin, H. (2023). Research status and prospects for the utilization of lead–zinc tailings as building materials. *Buildings, 13*(1), 150. https://doi.org/10.3390/buildings13010150
Li, X., Jiao, W., Rongbo, X., Chen, W., & Liu, W. (2017). Contaminated sites in China: Countermeasures of provincial governments. *Journal of Cleaner Production, 147*, 107. https://doi.org/10.1016/j.jclepro.2017.01.107
Li, X., Zhou, Y., & Zhang, J. (2021). Status and associated human health risk of zinc accumulation in agricultural soils across China. *Process Safety and Environmental Protection, 146*, 867–876. https://doi.org/10.1016/j.psep.2020.12.017
Liao, X., Li, Y., Miranda-Avilés, R., Zha, X., Anguiano, J. H. H., Moncada Sánchez, C. D., & Garzon, L. F. R. (2022). In situ remediation and ex situ treatment practices of arsenic-contaminated soil: An overview on recent advances. *Journal of Hazardous Materials Advances, 8*, 100157. https://doi.org/10.1016/j.hazadv.2022.100157
Lidman, J., Jonsson, M., & Berglund, Å. M. M. (2020). The effect of lead (Pb) and zinc (Zn) contamination on aquatic insect community composition and metamorphosis. *Science of the Total Environment, 734*, 139406. https://doi.org/10.1016/j.scitotenv.2020.139406
Lin, W., Xiao, T., Zhou, W., & Ning, Z. (2015). Pb, Zn, and cd distribution and migration at a historical zinc smelting site. *Polish Journal of Environmental Studies, 24*(2), 575–583. https://doi.org/10.15244/pjoes/26107
Liu, H., Wang, X., Zhang, B., Wang, W., Han, Z., Qinghua, C., & Liu, J. (2020). Concentration and distribution of lithium in catchment sediments of China: Conclusions from the China geochemical baselines project. *Journal of Geochemical Exploration, 215*, 106540. https://doi.org/10.1016/j.gexplo.2020.106540
Liu, H., Zhang, Y., Yang, J., Wang, H., Li, Y., Shi, Y., & Hu, W. (2021). Quantitative source apportionment, risk assessment and distribution of heavy metals in agricultural soils from southern Shandong peninsula of China. *Science of the Total Environment, 767*, 144879. https://doi.org/10.1016/j.scitotenv.2020.144879
Liu, K., Li, C., Tang, S., Shang, G., Yu, F., & Li, Y. (2020). Heavy metal concentration, potential ecological risk assessment and enzyme activity in soils affected by a lead-zinc tailing spill in Guangxi, China. *Chemosphere, 251*, 126415. https://doi.org/10.1016/j.chemosphere.2020.126415
Liu, L., Li, W., Song, W., & Guo, M. (2018). Remediation techniques for heavy metal-contaminated soils: Principles and applicability. *Science of the Total Environment, 633*, 206–219. https://doi.org/10.1016/j.scitotenv.2018.03.161
Liu, N., Zhao, J., Du, J., Hou, C., Zhou, X., Chen, J., & Zhang, Y. (2024). Non-phytoremediation and phytoremediation technologies of integrated remediation for water and soil heavy metal pollution: A comprehensive review. *Science of the Total Environment, 948*, 174237. https://doi.org/10.1016/j.scitotenv.2024.174237
Liu, Z., Bai, Y., Gao, J., & Li, J. (2023). Driving factors on accumulation of cadmium, lead, copper, zinc in agricultural soil and products of the North China plain. *Scientific Reports, 13*, 7429.
Mitra, S., Chakraborty, A. J., Tareq, A. M., Emran, T. B., Nainu, F., Khusro, A., & Simal-Gandara, J. (2022). Impact of heavy metals on the environment and human health: Novel therapeutic insights to counter the toxicity. *Journal of King Saud University Science, 34*(3), 101865. https://doi.org/10.1016/j.jksus.2022.101865
Moghimi Dehkordi, M., Pournuroz Nodeh, Z., Soleimani Dehkordi, K., Salmanvandi, H., Rasouli Khorjestan, R., & Ghaffarzadeh, M. (2024). Soil, air, and water pollution from mining and industrial activities: Sources of pollution, environmental impacts, and prevention and control methods. *Results in Engineering, 23*, 102729. https://doi.org/10.1016/j.rineng.2024.102729
Mohamed, E. S., Jalhoum, M. E. M., Hendawy, E., El-Adly, A. M., Nawar, S., Rebouh, N. Y., & Shokr, M. S. (2024). Geospatial evaluation and bioremediation of heavy metal-contaminated soils in arid zones. *Frontiers in Environmental Science, 12*, 1381409. https://doi.org/10.3389/fenvs.2024.1381409
Mosa, A., Taha, A. A., & Elsaeid, M. (2021). In-situ and ex-situ remediation of potentially toxic elements by humic acid extracted from different feedstocks: Experimental observations on a contaminated soil subjected to long-term irrigation with sewage effluents. *Environmental Technology & Innovation, 23*, 101599. https://doi.org/10.1016/j.eti.2021.101599
Negahban, S., & Mokarram, M. (2021). Potential ecological risk assessment of Ni, Cu, Zn, Cd, and Pb in roadside soils. *Earth and Space Science, 8*(4), e2020EA001120. https://doi.org/10.1029/2020EA001120
Nie, J., Wang, Q.-M., Han, L.-J., & Li, J.-S. (2024). Synergistic remediation strategies for soil contaminated with compound heavy metals and organic pollutants. *Journal of Environmental Chemical Engineering, 12*(4), 113145. https://doi.org/10.1016/j.jece.2024.113145

Nikolova, R., Boteva, S., Kenarova, A., Dinev, N., & Radeva, G. (2023). Enzyme activities in soils under heavy metal pollution: A case study from the surroundings of a non-ferrous metal plant in Bulgaria. *Biotechnology & Biotechnological Equipment, 37*(1), 49–57. https://doi.org/10.1080/13102818.2022.2149348

Noulas, C., Tziouvalekas, M., & Karyotis, T. (2018). Zinc in soils, water and food crops. *Journal of Trace Elements in Medicine and Biology, 49*, 252–260. https://doi.org/10.1016/j.jtemb.2018.02.009

Nriagu, J. (2019). Environmental pollution and human health in ancient times. In J. Nriagu (Ed.), *Encyclopedia of environmental health* (2nd ed., pp. 598–614). Elsevier.

Nyiramigisha, P., Komariah, & Sajidan. (2021). Harmful impacts of heavy metal contamination in the soil and crops grown around dumpsites. *Reviews in Agricultural Science, 9*, 271–282. https://doi.org/10.7831/ras.9.0_271

Okenmuo, F. C., & Ewemoje, T. A. (2023). Estimation of soil water erosion using RUSLE, GIS, and remote sensing in Obibia River watershed, Anambra, Nigeria. *DYSONA Applied Science, 4*(1), 6–14. https://doi.org/10.30493/das.2022.349144

Padhye, L. P., Srivastava, P., Jasemizad, T., Bolan, S., Hou, D., Shaheen, S. M., & Bolan, N. (2023). Contaminant containment for sustainable remediation of persistent contaminants in soil and groundwater. *Journal of Hazardous Materials, 455*, 131575. https://doi.org/10.1016/j.jhazmat.2023.131575

Paul, O., Jasu, A., Lahiri, D., Nag, M., & Ray, R. (2021). In situ and ex situ bioremediation of heavy metals: The present scenario. *Journal of Environmental Engineering and Landscape Management, 29*, 454–469. https://doi.org/10.3846/jeelm.2021.15447

Peng, Y., & Yu, G. I. (2024). Assessment of heavy metal pollution on agricultural land in Chengdu city under different anthropogenic pressures based on APCS-MLR modelling. *Ecological Indicators, 165*, 112183. https://doi.org/10.1016/j.ecolind.2024.112183

Pizarro Barraza, F., Thiyagarajan, D., Ramadoss, A., Manikandan, V. S., Dhanabalan, S. S., Abarzúa, C. V., & Thirumurugan, A. (2024). Unlocking the potential: Mining tailings as a source of sustainable nanomaterials. *Renewable and Sustainable Energy Reviews, 202*, 114665. https://doi.org/10.1016/j.rser.2024.114665

Poloko, N. (2019). Physical separation methods, part 1: A review. *IOP Conference Series: Materials Science and Engineering, 641*, 012023. https://doi.org/10.1088/1757-899X/641/1/012023

Priya, A. A.-O., Muruganandam, M., Ali, S. A.-O., & Kornaros, M. A.-O. (2023). Clean-up of heavy metals from contaminated soil by phytoremediation: A multidisciplinary and eco-friendly approach. *Toxics*. https://doi.org/10.3390/toxics11050422

Qi, M., Wu, Y. A.-O., Zhang, S., Li, G., & An, T. A.-O. (2023). Pollution profiles, source identification and health risk assessment of heavy metals in soil near a non-ferrous metal smelting plant. *International Journal of Environmental Research and Public Health*. https://doi.org/10.3390/ijerph20021004

Qin, Y., & Tao, Y. (2022). Pollution status of heavy metals and metalloids in Chinese lakes: Distribution, bioaccumulation and risk assessment. *Ecotoxicology and Environmental Safety, 248*, 114293. https://doi.org/10.1016/j.ecoenv.2022.114293

Rashid, A., Schutte, B. J., Ulery, A., Deyholos, M. K., Sanogo, S., Lehnhoff, E. A., & Beck, L. (2023). Heavy metal contamination in agricultural soil: Environmental pollutants affecting crop health. *Agronomy, 13*(6), 1521. https://doi.org/10.3390/agronomy13061521

Rita, E., Chizoo, E., & Cyril, U. S. (2021). Sustaining COVID-19 pandemic lockdown era air pollution impact through utilization of more renewable energy resources. *Heliyon, 7*(7), e07455. https://doi.org/10.1016/j.heliyon.2021.e07455

Riyam, N. K., Amal, I. H., & Hosam, M. S. (2022). Heavy metal's environmental impact. In M. S. Hosam & I. H. Amal (Eds.), *Environmental impact and remediation of heavy metals* (p. 1). IntechOpen.

Rouhani, A., Hejcman, M., & Trögl, J. (2024). A review of soil pollution by potentially toxic elements and remediation strategies in copper mining areas in Iran. *International Journal of Environmental Science and Technology*. https://doi.org/10.1007/s13762-024-05800-8

Selvi, A., Rajasekar, A., Theerthagiri, J., Ananthaselvam, A., Sathishkumar, K., Madhavan, J., & Rahman, P. K. S. M. (2019). Integrated remediation processes toward heavy metal removal/recovery from various environments-a review. *Frontiers in Environmental Science, 7*, 66. https://doi.org/10.3389/fenvs.2019.00066

Shen, F., Liao, R., Ali, A., Mahar, A., Guo, D., Li, R., & Zhang, Z. (2017). Spatial distribution and risk assessment of heavy metals in soil near a Pb/Zn smelter in Feng County, China. *Ecotoxicology and Environmental Safety, 139*, 254–262. https://doi.org/10.1016/j.ecoenv.2017.01.044

Shifaw, E. (2018). Review of heavy metals pollution in China in agricultural and urban soils. *Journal of Health & Pollution, 8*(18), 180607.

Shu, W., Ye, Z., Zhang, Z., Lan, C. Y., & Wong, M. (2005). Natural colonization of plants on five lead/zinc mine tailings in southern China. *Restoration Ecology, 13*, 49–60. https://doi.org/10.1111/j.1526-100X.2005.00007.x

Souza, L., Pomarolli, L., & Veiga, M. (2020). From classic methodologies to application of nanomaterials for soil remediation: An integrated view of methods for decontamination of toxic metal(oid)s. *Environmental Science and Pollution Research, 27*, 10205–10227. https://doi.org/10.1007/s11356-020-08032-8

Stewart, A. A.-O. (2020). Mining is bad for health: A voyage of discovery. *Environmental Geochemistry and Health, 42*, 1153–1165.

Sun, R., Gao, Y., & Yang, Y. (2022). Leaching of heavy metals from lead-zinc mine tailings and the subsequent migration and transformation characteristics in paddy soil. *Chemosphere, 291*, 132792. https://doi.org/10.1016/j.chemosphere.2021.132792

Tang, Y., Bi, X., Fayek, M., Stuart, F. M., Wu, L., Jiang, G., & Liang, F. (2017). Genesis of the Jinding Zn-Pb deposit, Northwest Yunnan Province, China: Constraints from rare earth elements and noble gas isotopes. *Ore Geology Reviews, 90*, 970–986. https://doi.org/10.1016/j.oregeorev.2017.04.024

Tchounwou, P. B., Yedjou, C. G., Patlolla, A. K., & Sutton, D. J. (2012). Heavy metal toxicity and the environment. *Experientia Supplementum, 101*, 133–164.

Thakur, A., Thakur, P., Rana, A., & Thakur, I. (2024). Policy and regulatory aspects for the remediation of heavy metals contaminated soils from different countries. In *Bio-organic amendments for heavy metal remediation* (pp. 447–457). Elsevier.

Wang, L., Yang, D., Li, Z., Fu, Y., Liu, X., Brookes, P. C., & Xu, J. (2019). A comprehensive mitigation strategy for heavy metal contamination of farmland around mining areas—Screening of low accumulated cultivars, soil remediation and risk assessment. *Environmental Pollution, 245*, 820–828. https://doi.org/10.1016/j.envpol.2018.11.062

Wang, X., Han, Z., Wang, W., Zhang, B., Wu, H., Nie, L., & Liu, Q. (2019). Continental-scale geochemical survey of lead (Pb) in mainland China's pedosphere: Concentration, spatial distribution and influences. *Applied Geochemistry, 100*, 55–63. https://doi.org/10.1016/j.apgeochem.2018.11.003

Wang, Y.-B., Liu, C.-W., & Wang, S.-W. (2015). Characterization of heavy-metal-contaminated sediment by using unsupervised multivariate techniques and health risk assessment. *Ecotoxicology and Environmental Safety, 113*, 469–476. https://doi.org/10.1016/j.ecoenv.2014.12.036

Wuana, R. A., & Okieimen, F. E. (2011). Heavy metals in contaminated soils: A review of sources, chemistry, risks and best available strategies for remediation. *International Scholarly Research Notices, 2011*(1), 402647. https://doi.org/10.5402/2011/402647

Xi, C. Z., Dai, T. G., & Huang, D. Y. (2008). Distribution and pollution assessments of heavy metals in soils in Zhuzhou, Hunan. *Geology in China, 35*, 524–530.

Xia, W., Wang, R., Zhu, B., Rudstam, L., Liu, Y., Xu, Y., & Chen, Y. (2020). Heavy metal gradients from rural to urban lakes in Central China. *Ecological Processes, 9*, 47. https://doi.org/10.1186/s13717-020-00251-8

Yao, W., Xie, X., Zhao, P., & Bai, J. (2014). Global scale geochemical mapping program — Contributions from China. *Journal of Geochemical Exploration, 139*, 9–20. https://doi.org/10.1016/j.gexplo.2013.09.007

Yu, R., He, L., Cai, R., Li, B., Li, Z., & Yang, K. (2017). Heavy metal pollution and health risk in China. *Global Health Journal, 1*(1), 47–55. https://doi.org/10.1016/S2414-6447(19)30059-4

Yu, Y. A.-O., Luo, H., He, L., Liu, W., Xu, R., Zhang, L., & Wei, F. (2020). Level, source, and spatial distribution of potentially toxic elements in agricultural soil of typical mining areas in Xiangjiang River basin, Hunan Province. *International Journal of Environmental Research and Public Health*. https://doi.org/10.3390/ijerph17165793

Zang, Z., Li, Y., Li, H., Guo, Z., & Zhang, R. (2020). Spatiotemporal variation and pollution assessment of Pb/Zn from smelting activities in China. *International Journal of Environmental Research and Public Health, 17*, 1968. https://doi.org/10.3390/ijerph17061968

Zeng, J., Li, C., Wang, J., Tang, L., Wu, C., & Xue, S. (2022). Pollution simulation and remediation strategy of a zinc smelting site based on multi-source information. *Journal of Hazardous Materials, 433*, 128774. https://doi.org/10.1016/j.jhazmat.2022.128774

Zhang, B., Yao, J., & Lee, H.-J. (2022). Economic impacts and challenges of Chinese mining industry: An input–output analysis. *Frontiers in Energy Research, 10*, 784709. https://doi.org/10.3389/fenrg.2022.784709

Zhang, S., Zeng, X., Sun, P., & Ni, T. (2023). Ecological risk characteristics of sediment-bound heavy metals in large shallow lakes for aquatic organisms: The case of Taihu Lake, China. *Journal of Environmental Management, 342*, 118253. https://doi.org/10.1016/j.jenvman.2023.118253

Zhang, X., Yang, L., Li, Y., Li, H., Wang, W., & Ye, B. (2011). Impacts of lead/zinc mining and smelting on the environment and human health in China. *Environmental Monitoring and Assessment, 184*, 2261–2273. https://doi.org/10.1007/s10661-011-2115-6

Zheng, H., Kong, S., Chen, N., Yan, Y., Liu, D., Zhu, B., & Qi, S. (2020). Significant changes in the chemical compositions and sources of PM2.5 in Wuhan since the city lockdown as COVID-19. *Science of the Total Environment, 739*, 140000. https://doi.org/10.1016/j.scitotenv.2020.140000

Zheng, X., Lin, H., Du, D., Li, G., Alam, O., Cheng, Z., & Li, J. (2024). Remediation of heavy metals polluted soil environment: A critical review on biological approaches. *Ecotoxicology and Environmental Safety, 284*, 116883. https://doi.org/10.1016/j.ecoenv.2024.116883

Zhou, J.-X., Xiang, Z.-Z., Zhou, M.-F., Feng, Y.-X., Luo, K., Huang, Z.-L., & Wu, T. (2018). The giant upper Yangtze Pb–Zn province in SW China: Reviews, new advances and a new genetic model. *Journal of Asian Earth Sciences, 154*, 280–315. https://doi.org/10.1016/j.jseaes.2017.12.032

Žibret, G., Gosar, M., Miler, M., & Alijagić, J. (2018). Impacts of mining and smelting activities on environment and landscape degradation—Slovenian case studies. *Land Degradation & Development, 29*(12), 4457–4470. https://doi.org/10.1002/ldr.3198

Citizens' Attitudes Toward Wildlife: Objective Choice or Subjective Wishes

Konstantin Zakharov, Mikhail Lomskov, and Alexander Konovalov

Abstract

It is generally recognized in modern urbanism that the environment plays a significant role in the quality of life of citizens. Many researchers place a heavy emphasis on the environmental, social, economic, and systemic functions of so-called green spaces in large cities. Meanwhile, in practice, natural territories are often perceived only from the perspective of managing recreational activities. The state and the form of nature in the city remain the subject of debate. To develop the right approach, a fundamental question must be answered: What should "wild" nature be like in a metropolis. A review of the modern literature demonstrates that city dwellers prefer to relax in natural or nature-like areas that have a natural vegetation structure and limited elements of urban landscaped parks. Given the choice, most people prefer areas with a biotope having a mosaic structure with complex and diverse arrays of plants and other vegetation. However, preferences vary across countries depending on education, culture, traditions, and social status. Attitudes toward nature in cities fluctuate constantly and can shift along individual experiences and urban milieus. A particularly significant issue in modern cities is children's limited contact with nature, and over time, each generation finds itself living in an increasingly attenuated natural environment.

Keywords

City · Citizens · Landscape beautification · Psychological · Well-being · Landscape · Recreation · Risk · Wildlife

Introduction

Natural and green planted areas in cities are crucial for the ecological well-being of cities and in many respects determine the quality of life for a significant part of the population in the modern world. A wide range of research in various countries has focused on this issue and shown the key role of wildlife and green spaces in reducing atmospheric pollution in urban environments (McPherson & Simpson, 1998; Nowak, 1993), reducing noise pollution (Fang & Ling, 2005; Gidlöf-Gunnarsson & Öhrström, 2007), improving the microclimate (Gómez et al., 2004; Lafortezza et al., 2009; McPherson, 1990), and performing many other environmental protection functions. For a review of

K. Zakharov (✉) · M. Lomskov · A. Konovalov
Department of Zoology, Ecology and Protection of Nature named after A. G. Bannikov, Faculty of Biotechnology and Ecology, Moscow State Academy of Veterinary Medicine and Biotechnology named after K. I. Skryabin, Moscow, Russian Federation

M. M. Anwar, A. Aziz (eds.), *Sustainable Landscape for Urban Resilience*, Sustainable Development Goals Series, https://doi.org/10.1007/978-3-031-99836-2_12

the literature, see, for example, Carrus et al. (2013). A direct positive relationship has been established between the health of citizens and their proximity to nature (de Vries et al., 2003; Maas et al., 2006); for a review of the literature on this subject, see Hartig et al. (1991) and Tzoulas et al. (2007). In fact, nature constitutes the main factor supporting human well-being in urban environments (Chiesura, 2004; IUCN, 1991). For this reason, nature preserves and green spaces (which are often collectively covered by the term urban green spaces in international contexts) have become a priority in the modern world (Adinolfi et al., 2014), and not only for specialists in nature conservation, biology, and geography but also for economists, sociologists, psychologists, and others (Pregio et al., 2008).

Until recently, nature in the city has been mainly looked upon as a decorative and aesthetic element, but now the environmental, social, economic, and systemic functions of green spaces (which are most important for large cities) are taking on primary importance (Dobbs et al., 2011). Many studies have revealed the role of urban forests and parks in reducing stress and improving people's mental health (Berman et al., 2008; Conway, 2000; Fuller et al., 2007; Hartig et al., 2014; Kaplan, 1995). In fact, urban nature plays a strategic role in safeguarding the quality of life in our rapidly urbanizing society (Chiesura, 2004; Standish et al., 2013). At the same time, in practice, in an urban economy, green spaces are still often regarded only as places where people can relax, in direct contradiction of their leading role in environmental protection and nature conservation (Gobster, 2001). Human needs are normally appealed to as justification for the necessity to do landscaping of urban forests and other nature preserves, that is, for their transformation into urban parks adapted for recreational activities (Parsons, 1995; Yang et al., 2014). The situation becomes even more complicated through the desire to give a city a well-groomed appearance, to please the most diverse groups of citizens, and to adapt nature to a wide range of recreational activities. In many cities, urban landscaping of city forests and river valleys has particularly expanded: Some citizens encourage such activities, and some protest against it. This is an already classic debate between two opposing points of view on wildlife in cities (Daniel, 2001; Yang et al., 2014). One of them, the traditional, or anthropocentric, perspective, is based on the aesthetic preferences of inhabitants and asserts that the appearance and fate of urban natural areas should be determined by inhabitants themselves, since a person may feel uncomfortable in a natural environment (Bixler & Floyd, 1997). Proponents of the approach believe that this enables the most favorable environment for fostering safety, comfort, and outer visual appeal (Appleyard, 1979; Buchecker et al., 2003; Kelsey, 2008; Nohl, 1987). Urban nature is rather perceived as an object of transformation, with the focus on the design and functionality of a given space (Cosgrov, 2008; Ellison, 2013; Jellicoe & Jellicoe, 1995). The opposite approach began to take shape in the second half of the twentieth century, with the beginning of the explosive urbanization of natural landscapes. It can be described as "biocentric," and its proponents argue that even in urban areas, natural landscapes with natural flora and fauna are most valuable and outwardly attractive (Daniel, 2001). Such "ecological aesthetics" is discussed in the work by the famous environmentalist Aldo Leopold (Gobster, 1995). Environmental knowledge has penetrated landscape architecture largely through McKinney (2002) work; his "Design with Nature" is considered significant in this respect (Herrington, 2010).

These opposing points of view continue to trigger heated debate: Conservationists justify the need to preserve wildlife in cities (McKinney, 2002), while their opponents criticize the "green philosophy" as being environmental determinism and antihuman (Monzingo, 1997) or empty sophistry that impedes urban development (Nadenicek & Hastings, 2000). Discussions of practical issues often drift into the field of philosophy (Lothian, 1999; Zhang, 2013). In recent years, more and more suggestions have been made to integrate the two opposing viewpoints (Gobster et al., 2007; Yang et al., 2014).

Material and Methods

To illustrate city dwellers' interest in these issues, we used Google Trends in the period 2004–2025 with the condition "the whole world." We used three queries searching "green areas in city," "urban green space," and "green infrastructure." In this way, we tried to explain why modern city dwellers pay attention to wildlife and which factors influence them in this connection. To understand the interest in landscape beautification, we used the terms "beautification in city" and "urban design." MS Excel was used to process the data. Google Scholar was used to search articles in the period 1990–2020. We used our three queries again, and from among the answers we chose the most popular scientific works with a predominance of review articles and sources from different countries from 1990.

Results

Google Trends use has yielded unequal results, with a predominance of the term "green infrastructure," in particular from 2020. Our results are shown in Fig. 1. Each diagram contains a trend line with the coefficient $R^2 > 0.4$. Regressions equals are shown in Table 1. Trend lines confirm the increasing attention being given to our topic.

The query "beautification in city" did not yield results sufficient to construct a diagram, so we used the topic "urban design" suggested by Google Trends, which has a clearly declining trend (Fig. 2).

Review of the scientific literature showed that, besides city dwellers' needs, several other important issues arise, which are discussed in what follows.

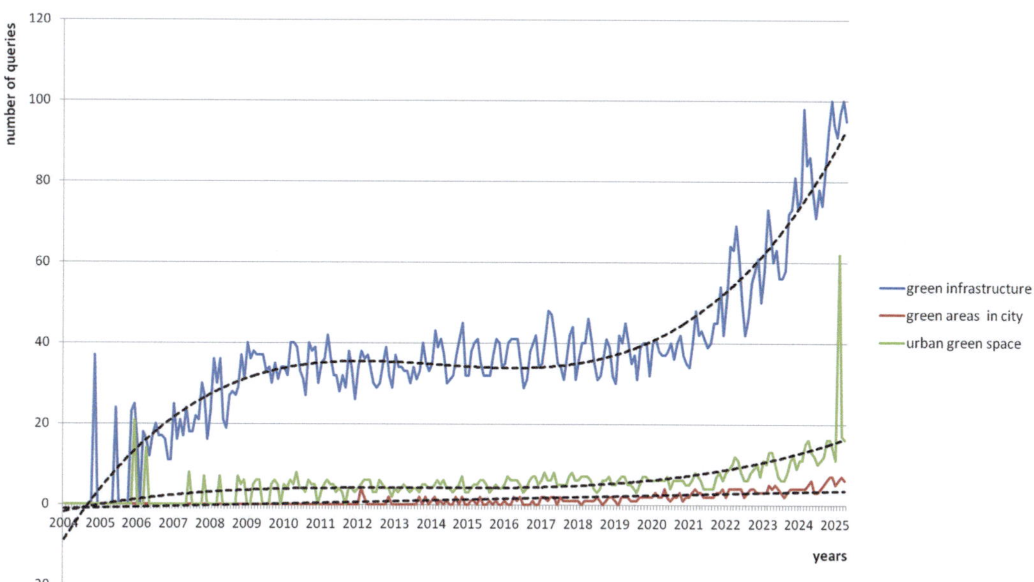

Fig. 1 Results of query mining from Google Trends. Dotted lines denote trends

Table 1 Regression equals of trend line for queries in Google Trends

Query	Regression equal	R^2
Urban design	$y = 0.0013x^2 - 0.5466x + 85.122$	0.93
Green infrastructure	$y = 3E\text{-}05x^3 - 0.0101x^2 + 1.1892x - 9.9355$	0.89
Green areas in city	$y = 0.0179x - 1.1031$	0.64
Urban green space	$y = 4E\text{-}06x^3 - 0.0015x^2 + 0.1638x - 1.9633$	0.47

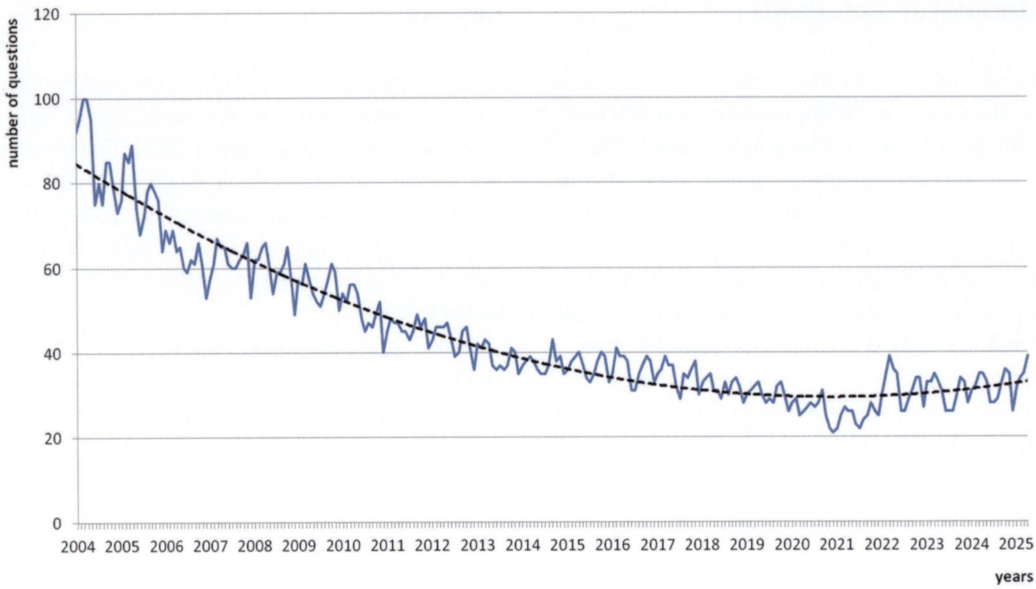

Fig. 2 The number of queries for "urban design" (blue line) and trend (dotted line)

Modern Inhabitants' Need to Interact with Wildlife

A detailed review and analysis of various opinions about the place and form of nature in the city is quite complicated and too vast to be covered in a single chapter like this. Meanwhile, amid active disputes, the fundamental question remains sidelined, though it should be featured prominently in decision making: What does a person really need, what kind of natural environment is optimal for the population as a whole, for people of various ages, social statuses, and personal preferences, in other words, *what form of nature do people want*? The aim of the review was to search for answers to these questions.

To start with, it is necessary to determine whether residents of different categories have any common needs. This goal is discussed in the generalizing work of Matsuoka and Kaplan (2008), so we will dwell on it in greater detail. This work analyzed 90 publications in the scientific journal *Landscape and Urban Planning* for the period from 1991 to 2006. Two groups of needs shared by all residents were established that can be conventionally called "natural" and "social." We list them in decreasing order of the number of references in works of various authors.

Natural Needs

These are the basic needs of residents discussed in 92% of works on the topic.

Regular contact with nature: This is the most important need of modern city dwellers. It includes various forms of communing with nature and its elements, the opportunity to see nature in a city and be in contact with it. Regular contact with wildlife is directly associated with improving mental and physical health in humans and with a feeling of satisfaction and ease with the environment. The papers that deal with this topic show the importance of communication between citizens and nature in both large areas (e.g., in forests) and small gardens, squares, and even green courtyards. The so-called green atmosphere (or "sense of nature") is recognized as the most important factor contributing to citizens' more favorably assessing their area of residence. The economic benefits of urban nature are also shown. In particular, people's desire to live closer to nature significantly increases housing prices and hotel room prices (Crow et al., 2006).

Aesthetic preferences: This category includes such features as an attractive natural landscape, cleanliness, and lack of noise, for example. Residents clearly prefer to see natural landscapes near their homes rather than urbanized areas with pockets of greenery.

The need for recreation: Daily outdoor recreation primarily includes a wide variety of recreational activities (e.g., walks in nature, sports, outdoor games).

Social Needs

This group of publications looks at enhancing the potential for residents to be in nature if green spaces are located near their home. In general, the works analyzed considered this need less often (56%) than the need to communicate with nature.

Social interaction and privacy: The works discussing this problem show that residents' social interactions can be improved through a nature-like appearance of green spaces. In this case, they contribute to strengthening contacts among various social groups and neighbors. Natural landscapes do play a key role in promoting social interactions among people. Along with this, such landscapes function as shelters from the hustle and bustle of city life and foster the kind of "personal space" that is especially needed by residents of large cities.

Citizens' involvement in creating the image of a city: The works on this topic manifest the desire of many residents to improve the appearance of their city. Moreover, designers, managers, and social researchers try to take this need into account. It is typical not only of democratic countries but also of nondemocratic ones, for example, Jordan and Saudi Arabia.

A sense of shared identity: The loss of identity by residents in large cities is evident throughout the world. There are a few causes of this, among which the faceless appearance of many modern cities is prominent. For example, researchers in the Middle East point to the misguided urban development projects in the region that seek to generate economic benefits rather than to meet residents' needs (Abu-Ghazzeh, 1996) and to provide nonselective zoning of urban areas (Al-Hathloul & Mughal, 1999). In the United States, a number of researchers have discussed the loss of identity in cities (Lucy & Phillips, 1997); they assert that enhancing the identity of a particular city through the preservation and restoration of the natural environment may contribute to a feeling of community among residents. Other researchers show that green spaces in a city, especially nature preserves, increase the identity of a society (Stewart et al., 2004). Thus, the analyzed publications primarily emphasize the importance of the proximity of places of residence to green spaces. Although there are different reasons account for the need to communicate with nature in various regions of the world, they are quite similar, despite differences in the way of life of people, cultures, and political systems (Matsuoka & Kaplan, 2008).

There are empirical works along with general studies that directly raise the question of reasons for visiting parks and other green spaces. One of the most frequently cited is a study by Chiesura (2004) in Amsterdam. A survey of park visitors revealed the reasons residents have for being in nature. The researcher concludes that urban nature satisfies many of the social and psychological needs of people that make it a most valuable resource and a key component of the sustainability of a city (Chiesura, 2004). Similar results were obtained in studies on the motivations for visiting green spaces in British cities (Burgess et al., 1988; Irvine et al., 2013), Turkey (Oguz, 2000), China (Jim & Chen, 2006), and Iran (Abkar et al., 2010). Thus, the results of specialized studies assume that citizens visit parks and other green spaces primarily for the purpose of being outdoors. This conclusion allows us to return to the main question of the present review of the literature: What should nature be like in a city in order to satisfy the needs of its inhabitants?

Many researchers study urban recreation and explore its effect on the physical and psychological state of individuals, while at the same time analyzing their perceptions of the environment (Carrus et al., 2013). A clear relationship has been established between perceptions of environmental components and the enjoyment of leisure activities (Hartig, 2004; Kaplan, 1995). Most researchers who compare leisure activities in either urban or natural surroundings show that people unequivocally prefer a natural environment without elements of urbanization. For a review of the literature, see Carrus et al. (2013).

The Key Elements of Nature for Residents

Here are some examples. Residents' perceptions of natural vegetation and formally planted green spaces in the city of Bari, Italy, were investigated. A comparison was made between a garden square, a city park, forest plantations, a botanical garden, and natural suburban forests. All areas were selected by biologists according to the degree of increase in their natural qualities: biological diversity, canopy density, and origin (natural or manmade), as well as accessibility and the attractive appearance of the landscape. Garden squares have a significant proportion of paved surfaces and relatively poorly developed vegetation. City parks have more highly developed vegetation, but limited biological diversity. Forest plantations are characterized by a high degree of biological diversity but have a homogeneous tree layer. Botanical gardens have a high variety of plants and a limited number of urbanization elements. Large suburban natural areas have maximum vegetation cover and a high variety of flora and fauna.

The results of the study were as follows. With the exception of the garden square, all territories are more or less attractive for recreation, and the area having the most natural elements being the most attractive. The study clearly shows a correspondence between the degree of recreational attractiveness of a territory and its natural features. In practical terms, this work demonstrates that the appeal of city green spaces for recreational use can be increased through a comprehensive strengthening of their natural qualities (Carrus et al., 2013).

A comparison of natural and green areas that are fundamentally different in their features helped to establish that vacationers prefer to spend time in a natural environment. Natural vegetation holds the greatest recreational appeal and is more often chosen by city dwellers as a place to unwind, also taking into account their aesthetic preferences. What exactly attracts people to nature and what is most important for them when it comes to relaxing? Back in the 1980s, the positive effects of nature on human health were researched by health workers. For example, Ulrich (1984, 1992) found that postoperative patients recovered faster and required less effort to relieve pain if they could see trees through a window rather than a brick wall. Other researchers came to the conclusion that the greatest psychological effect was achieved in natural areas with high biodiversity; moreover, visitors experienced differences in the variety of species of plants and animals in different areas. Studies in Sheffield (England) showed that the degree of psychological benefit positively correlates with the species abundance of plants and, to a lesser extent, of birds and butterflies. People are more careful evaluating the visible static component of nature, the diversity of plants and vegetation layers. The perception of the abundance of birds and butterflies is less pronounced. The psychological comfort of a person is associated with the diversity of habitats in a particular area and shows the key importance of a significant heterogeneity in natural vegetation for psychological well-being. Consequently, a created mosaic of biotopes, in contrast to homogeneous green spaces, improves residents' psychological well-being. When no consideration is given to the biological diversity of an area that is protected from development, this both reduces the area's recreational appeal and may completely depreciate it as a venue of leisure. At the same time, the preservation and strengthening of the natural qualities of such areas create new opportunities for residents to come into contact with nature, thereby increasing the psychological well-being of the population (Fuller et al., 2007).

Studies conducted in Flanders in northern Belgium, where the long history of development has resulted in a long-established cultural landscape, are indicative. In recent decades, there has been an increased demand for recreation in a natural environment, and the townspeople are primarily drawn to forests. In the European Union, such needs are recognized and reflected in the concept of development and maintenance of urban forestry (COST, 1997). The preferences and expectations of visitors and their perceptions of the forest were determined in order to establish the most attractive forms of forests (Rydberg & Falck, 2000). Studies were conducted in the forest complex Heverlee-Meerdaal (Flanders). It was established that most visitors came to the forest for recreational walks, jogging, and cycling. There is a negative correlation between visits and the remoteness of forests from people's homes. The main conclusion of the researchers was that visitors preferred heterogeneous forests with a complex composition and structure of vegetation, which should be taken into account in the practice of forestry (Roovers et al., 2002).

Similar conclusions were reached in a study carried out in Granada in northeastern Spain. A statistical analysis showed a significant correlation between park attendance and the number of trees per unit area (Adinolfi et al., 2014). Research in India has also established another key factor for people visiting green spaces: the richness of species and the diversity of heterogeneously distributed biotopes (Nagendra & Gopal, 2010). A picturesque natural landscape has a positive effect on the recreational appeal of an area (Chen et al., 2009). Developed forest vegetation contributes to a comfortable microclimate, which is one of the factors that attract visitors. Many researchers have compared the comfort of recreational conditions in forests and beyond (Baris et al., 2009; Gómez et al., 2004; Lafortezza et al., 2009; McPherson & Simpson, 2003; Spronken-Smith & Oke, 1998).

All of these studies confirm a significant correspondence between the projective cover of woody vegetation and the number of visitors (Adinolfi et al., 2014). Unlike the forest environment, park-type green spaces with a simple structure attract few visitors, especially during hot times of the day (Georgi & Zafiriadis, 2006; Givoni, 1991). Studies in Granada showed the importance of the four main factors for recreation: the natural appearance of an area, high biodiversity, proximity to place of residence, and easy access, as well as some elements of landscaping within certain limits (Adinolfi et al., 2014). Similar results were obtained by many researchers in other countries, including China (Yu, 1995), Japan (Oku & Fukamachi, 2006; Ozguner & Kendle, 2006), Malaysia (Noralizawati, 2009), and Korea (Lee & Kim, 2015). The main conclusion to be drawn from the analysis of the cited literature is that citizens prefer close-to-home, natural, or nature-like areas for recreation that have a natural vegetation structure and limited elements of park landscaping. The results of the review may not seem to fit well with the already conventional urges to "improve on nature" and to "put it in order" that are usually used as the basis for landscaping and transforming natural areas into amusement parks with green plantations. What lies behind this contradiction?

The Changing Attention of Dwellers to Wildlife

The question of residents' preferences and their sympathies for certain landscapes in a city has been particularly actively studied in the last two to three decades (Carlson, 2006; Yang et al., 2014). An empirical study by Dearden (1984) is considered to be the first on this topic. The respondents were presented with color photographs showing three types of landscape (urban, suburban, or natural) and were asked to choose the most attractive. The results showed significant differences in the preferences exhibited by the participants from a local environmental movement and other groups of respondents; by people familiar or unfamiliar with the local nature; and by residents of neighborhoods with sparse or dense buildings. Further, it was demonstrated that the aesthetic assessment of the landscape and other indicators were influenced by the age of the respondents and their place of residence (Tahvanainen et al., 2001; Van den Berg & Koole, 2006), educational attainments (Harvey, 1995; Howell & Laska, 1992; Steel et al., 1994), and the type of landscape (Junker & Bucheker, 2008). When studying the preferences of residents of Germany, Spain, and Chile, differences were established depending on cultural levels, access to natural spaces, and national traditions; social status also played an important role (Pregio et al., 2008). Moreover, the preferences of residents are affected by complex socioeconomic, cultural, and biophysical interactions (Barlam & Dragićević, 2005; Fraser & Kenney, 2000), as well as by the surrounding world as a whole and the way it is perceived (Zhang et al., 2010). Attitudes toward nature are both a mental phenomenon and a consequence of individual experience and cultural and social factors; environmental perception is rather mutable and variable (Kaymaz, 2012). Thus, the value of nature in the eyes of the next generation of city dwellers usually decreases: as urbanization intensifies, the opportunities for coming into contact with nature diminish, including in childhood. Modern city dwellers simply cannot imagine what the natural environment should be like, and they often welcome the creation of familiar green spaces. This is recognized as a very dangerous phenomenon, as now there is an increasing need for a healthy environment, although many fail to acknowledge it (Miller, 2005).

Some Psychological Theories to Explain People's Aesthetic Preferences

Various theories have been developed in an attempt to explain aesthetic preferences, including sympathies for a transformed landscape; see the reviews by Kaymaz (2012) and Zhang (2013). Psychologists Rachel and Stephen Kaplan, well known in nature conservation, consider landscapes with homogeneous and thick vegetation cover and other visual obstacles to be less preferable. In such an environment, people may feel insecure (Kaplan & Kaplan, 1978). Appleton (1975) looked at aesthetic preferences as a genotypic phenomenon: a semi-open landscape with separated large trees (which are beloved by many) resembles the African savannah. Nassauer (1995) explained the urge to improve nature by the desire of people to see an area as managed and "not abandoned." Perhaps this is due to the centuries-old use and transformation of nature for agricultural purposes (Breuste, 2004), as well as the well-established rejection of ravines, swamps, floodplain forests, and other naturally valuable areas that are of little interest to peasants (Duhme & Pauleit, 1992). However, it should be repeated that modern citizens, when given a choice among recreation spots, prefer a natural environment to a planted one, and high aesthetic assessments of a landscape are directly related to its biological diversity (Junker & Bucheker, 2008; Nassauer, 2004).

Thus, along with the study of the impact of the natural environment on humans, the focus of attention of many researchers is on the preferences of residents for particular landscapes. The question of "what nature is better or worse in a city" has long ventured beyond the bounds of emotional debate and acquired clear and precise answers. Complex vegetation and high biological diversity are recognized as attracting people and as being perceived as the most important indicators of natural areas and green spaces. Apart from providing ecosystem services, nature also plays an important social role, especially in large, featureless cities with dense and repetitive development.

The results of studying the aesthetic assessment of landscapes indicate the need for a very cautious attitude toward the urge to "put nature in order" since they are influenced by many factors that do not always reflect people's true needs. We should distinguish between the objective ecological state of an area and subjective aesthetic preferences (Junker & Bucheker, 2008; Nassauer, 2004). Allegations of the "impoverished" state of natural territories and the need for their "development" indicate only a superficial familiarity with this issue. Managers, designers, landscape architects, and others who are responsible for the state of natural and green areas in a city should be guided in their decision-making by modern scientific knowledge in the field of urban ecology, rather than personal ideas and preferences. In a modern city, nature can no longer be seen as something insignificant and marginal; it requires a very balanced approach, since decisions directly associated with it affect the quality of life of residents.

Conclusion

Overall, our review has revealed a contradiction between two important issues, the needs of residents and ideas about ideal nature in cities. On the one hand, residents choose green areas with a natural appearance when it comes to relaxing and avoid areas that have been beautified. On the other hand, the increasing processes of urbanization and alterations of urban environments, which alter residents' awareness, and those residents increasingly take artificial vegetation cover as given. In particular, new generations of residents are growing up in modern settings in the shadows of skyscrapers and have little contact with nature, which is undoubtedly a big problem for modern society. In this study, our Internet-based investigation uncovered a growing interest in nature among city dwellers, but exactly what that interest reveals is unclear because citizens may simply accept artificial vegetation cover as the normal state of urban green spaces. Residents' awareness of the state of nature is a difficult phenomenon to gauge because it changes more quickly than we can measure. Thus, study of the influence of the environment on city dwellers' perceptions represents a fruitful direction for future investigation.

References

Abkar, M., Schariff, M., Kamal, M., Mariapan, M., Maulan, S., & Sheybani, M. (2010). The role of urban green spaces in mood change. *Australian Journal of Basic and Applied Sciences, 4*, 5352–5361.

Abu-Ghazzeh, T. M. (1996). Reclaiming public space: The ecology of neighborhood open spaces in the town of Abu-Nuseir, Jordan. *Landscape and Urban Planning, 36*, 197–216.

Adinolfi, C., Suárez-Cáceres, G. P., & Cariñanos. (2014). Relation between visitors, behaviour and characteristics of green spaces in the city of Granada, South-Eastern Spain. *Urban-Forestry & Urban Greening, 3*, 534–542.

Al-Hathloul, S., & Mughal, M. A. (1999). Creating identity in new communities: Case studies from Saudi Arabia. *Landscape and Urban Planning, 44*, 199–218.

Appleton, J. (1975). *The experience of landscape*. Wiley.

Appleyard, D. (1979). The environmental and as a social symbol: Within a theory of environmental action and perception. *Journal of the American Planning Association, 45*(2), 143–153.

Baris, M. E., Satris, S., & Yargon, M. E. (2009). The contribution of trees and green spacesto the urban climate: The case of Ankara. *African Journal of Agricultural Research, 4*(9), 791–800.

Barlam, S., & Dragićević, S. (2005). Attitude toward urban green spaces: Integrating questionnaire survey and collaborative GIS techniques to improve attitude measurements. *Landscape and Urban Planning, 71*, 147–162.

Berman, M. G., Jonides, J., & Kaplan, S. (2008). The cognitive benefits of interacting with nature. *Psychological Science, 19*, 1207–1212.

Bixler, R. D., & Floyd, M. F. (1997). Nature is scary, disgusting and uncomfortable. *Environment and Behavior, 29*, 443–467.

Breuste, J. H. (2004). Decision making, planning and design for the conservation of indigenous vegetation within urban development. *Landscape and Urban Planning, 68*, 439–452.

Buchecker, M., Hunziker, M., & Felix, K. (2003). Participatory landscape development: Overcoming social barriers to public involvement. *Landscape and Urban Planning, 64*(1-2), 29–46.

Burgess, J., Harrison, D., & Limb, M. (1988). People, parks and urban green: A study of popular meanings and values for open spaces in the city. *Urban Studies, 25*, 455–573.

Carlson, A. (2006). The aesthetic appreciation of environmental architecture under different conceptions of environment. *The Journal of Aesthetics Education, 40*(4), 77–88.

Carrus, G., Lafortezza, R., Colangelo, G., Dentamaro, I., Scopelliti, M., & Sanesi, G. (2013). Relations between naturalness and perceived restorativeness of different urban green spaces. *Psychology, 4*(3), 227–244.

Chen, B., Adino, O. A., & Bao, Z. (2009). Assessment of aesthetic quality and multiple functions of urban green spaces for the users' perspective: The case of Hangzhou flower garden, China. *Landscape and Urban Planning, 93*, 76–82.

Chiesura, A. (2004). The role of urban parks for the sustainable city. *Landscape and Urban Planning, 68*, 129–138.

Conway, H. (2000). Parks and people: The social functions. In J. Woudstra & K. Fieldhouse (Eds.), *The regeneration of public parks*. Routledge.

Cosgrov, D. E. (2008). Seminar discussion. In R. Z. Delue & J. Elkins (Eds.), *Landscape theory* (p. 128). Routledge.

COST. (1997). *Memorandum of understanding for the implementation of a European concerted research action designed as COST action E12 "urban forests and trees"*. European Commission.

Crow, T., Brown, T., & DeYoung, R. (2006). The Riverside and Berwyn experience: Contrasts in landscape structure, perceptions of the urban landscape, and their effects on people. *Landscape and Urban Planning, 75*, 282–299.

Daniel, T. C. (2001). Whither scenic beauty? Visual landscape quality assessment in the 21st century. *Landscape and Urban Planning, 54*, 267–281.

de Vries, S., Verheij, R., Groenewegen, P., & Spreeuwenberg, P. (2003). Natural environments-healthy environments? An exploratory analysis of the relationship between greenspace and health. *Environment and Planning, 35*, 1717–1732.

Dearden, P. (1984). Factors influencing landscape preferences: An empirical investigation. *Landscape and Urban Planning, 11*, 293–306.

Dobbs, C., Escobedo, F. J., & Zipperer, W. C. (2011). A framework for developing urban forest ecosystem services and good indicators. *Landscape and Urban Planning, 99*, 196–206.

Duhme, F., & Pauleit, S. (1992). Naturschutzprogramm für München. Landschaftsökologisches Ramenkonzept. *Geographische Rundschau, 44*(10), 554–561.

Ellison, A. (2013). The suffocating embrace of landscape and picturesque conditioning of ecology. *Landscape Journal, 32*(1), 79–94.

Fang, C. E., & Ling, D. L. (2005). Guidance for noise reduction provided by tree belts. *Landscape and Urban Planning, 71*, 29–34.

Fraser, E. D. G., & Kenney, W. A. (2000). Culture background and landscape history as factors affecting perceptions of the urban forest. *Journal of Arboriculture, 26*(2), 106–112.

Fuller, R. A., Irvine, K. N., Devine-Wrigh, P., Warren, P. H., & Gaston, K. J. (2007). Psychological benefits of greenspace increase with biodiversity. *Biology Letters, 3*, 390–394.

Georgi, N. J., & Zafiriadis, K. (2006). The impact of park trees on microclimate in urban areas. *Urban Ecosystems, 9*, 195–209.

Gidlöf-Gunnarsson, A., & Öhrström, E. (2007). Noise and well-being in urban residential environments: The potential role of perceived availability to near green areas. *Landscape and Urban Planning, 83*, 115–126.

Givoni, B. (1991). Impact of planted areas on urban environmental quality: A review. *Atmospheric Environment Part B: Urban Atmosphere, 25*, 289–299.

Gobster, P. (1995). Aldo Leopold's ecological esthetic: Integrating esthetic and biodiversity values. *Journal of Forestry, 93*, 6–10.

Gobster, P. (2001). Vision of nature: Conflict and compatibility in urban park restoration. *Landscape and Urban Planning, 56*, 35–51.

Gobster, P. H., Nassauer, J. I., Daniel, T. S., & Fry, C. (2007). The shared landscapes: What does aesthetics have to do with ecology? *Landscape Ecology, 22*, 959–972.

Gómez, F., Gil, L., & Jabaloyes, J. (2004). Experimental investigation of the thermal com-fort in the city: Relationship with the green areas, interaction with the urban microclimate. *Building and Environment, 39*, 1077–1086.

Hartig, T. (2004). Restorative environments. In C. Spielberger (Ed.), *Encyclopedia of applied psychology* (pp. 273–279). Academic Press.

Hartig, T., Mang, M., & Evans, G. (1991). Restorative effects of natural environments experiences. *Environment and Behavior, 23*, 3–26.

Hartig, T., Mitchell, R., de Vries, S., & Frumkin, H. (2014). Nature and health. *Annual Review of Public Health, 35*, 207–228.

Harvey, T. (1995). An education 21 programme: Orienting environmental education towards sustainable development and capacity building for Rio. *The Environmentalist, 15*, 2002–2012.

Herrington, S. (2010). The nature of Ian McHarg's science. *Landscape Journal, 29*, 1–10.

Howell, S. E., & Laska, S. B. (1992). The changing face of the environmental coalition. *Environment and Behavior, 24*, 134–144.

Irvine, K. N., Warber, S. L., Devine-Wright, P., Gaston, K. J., & Int, J. (2013). Understanding urban green space as a health resource: A qualitative comparison of visit motivation and derived effects among park users in Sheffield, UK. *Environmental Research and Public Health, 10*, 417–442.

IUCN. (1991). *Caring for the earth: A strategy for sustainable living*. The World Conservation Union, Worldlife Fund for Nature.

Jellicoe, G., & Jellicoe, S. (1995). *The landscape of man: Shaping the environment from prehistory to the present day* (3rd ed.). Themes and Hudson.

Jim, C., & Chen, W. (2006). Recreation-amenity use and contingent valuation of urban greenspaces in Guangzhou, China. *Landscape and Urban Planning, 75*, 81–96.

Junker, B., & Bucheker, M. (2008). Aesthetics preferences versus ecological objectives in river restorations. *Landscape and Urban Planning, 85*, 141–154.

Kaplan, R., & Kaplan, S. H. (1978). *Environments for people*. Duxbure Press.

Kaplan, S. (1995). The restorative benefits of nature: Toward an integrative framework. *Journal of Environmental Psychology, 15*, 169–182.

Kaymaz, I. K. (2012). Landscape perception. In M. Ozyavuz (Ed.), *Landscape planning* (pp. 251–275). IntechOpen.

Kelsey, R. (2008). Seminar discussion. In R. Z. Delue & J. Elkins (Eds.), *Landscape theory* (p. 128). Routledge.

Lafortezza, R., Carrus, G., Sanesi, G., & Davies, C. (2009). Benefits and well-being perceived by people visiting green spaces in periods of heat stress. *Urban Forestry and Urban Greening, 8*, 97–108.

Lee, Y.-C., & Kim, K.-H. (2015). Attitudes of citizens towards urban parks and green spaces for urban sustainability: The case of Gyeongsan City, Republic of Korea. *Sustainability, 7*, 8240–8254.

Lothian, A. (1999). Landscape and the philosophy of aesthetics: Is the landscape quality inherent in the landscape or in the eye of the beholder? *Landscape and Urban Planning, 44*, 177–198.

Lucy, W. H., & Phillips, D. L. (1997). The post-suburban era comes to Richmond: City decline, suburban transition, and exurban growth. *Lands and Urban Plan, 36*, 259–275.

Maas, J., Verheij, R. A., Groenewegen, P. P., de Vries, S., & Spreeuwenberg, P. (2006). Green space, urbanity, and health: How strong is the relation? *Journal of Epidemiology and Community Health, 60*, 587–592.

Matsuoka, R. H., & Kaplan, R. (2008). People needs in the urban landscape: Analysis of landscape and urban planning contributions. *Landscape and Urban Planning, 84*, 7–19.

McKinney, M. L. (2002). Urbanization, biodiversity and conservation. *Bioscience, 52*(10), 883–890.

McPherson, E. G. (1990). Modeling residential landscape water and energy use to evaluate water conservation policies. *Landscape Journal, 9*, 122–134.

McPherson, E. G., & Simpson, J. R. (1998). Air pollutant uptake by Sacramento's urban forest. *Journal of Arboriculture, 24*, 224–234.

McPherson, E. G., & Simpson, J. R. (2003). Potential energy savings in buildings by an urban tree planting program in California. *Urban Forestry and Urban Greening, 2*, 73–86.

Miller, J. R. (2005). Biodiversity conservation and the extinction of experience. *Trends in Ecology and Evolution, 20*(8), 430–434.

Monzingo, L. A. (1997). The aesthetics of ecology design: Seeing science as culture. *Landscape Journal, 16*(1), 46–59.

Nadenicek, D. J., & Hastings, C. M. (2000). Environmental rhetoric, environmental sophism: The words and works of landscape architecture. In M. Concan (Ed.), *Environmentalism in landscape architecture*. Dumbarton Oaks.

Nagendra, H., & Gopal, D. (2010). The diversity, distribution, history and change in urban parks: Studies in Bangalore, India. *Urban Ecosystem, 14*, 211–223.

Nassauer, J. I. (1995). Messy ecosystems, orderly frames. *Landscape Journal, 14*(2), 161–170.

Nassauer, J. I. (2004). Monitoring the success of metropolitan wetland restorations: Culture sustainability and ecological function. *Wetlands, 24*, 756–765.

Nohl, W. (1987). The aesthetics of home separated gardens in Germany: Traces of participatory aesthetics. *Journal of Architectural and Planning Research, 4*(3), 212–227.

Noralizawati, M. (2009). *Public preferences towards naturalistic and designed landscape pattern in park area*. University of Technology.

Nowak, D. J. (1993). Atmospheric carbon-reduction by urban trees. *Journal of Environmental Management, 37*(3), 207–217.

Oguz, D. (2000). User surveys of Ankara's urban parks. *Landscape and Urban Planning, 52*, 165–171.

Oku, H., & Fukamachi, K. (2006). The differences in scenic perception of forest visitors through their attributes and recreational activity. *Landscape and Urban Planning, 75*, 34–42.

Ozguner, H., & Kendle, A. D. (2006). Public attitudes towards naturalistic versus designed landscapes in the city of Sheffield (UK). *Landscape and Urban Planning, 74*, 139–157.

Parsons, R. (1995). Conflict between ecology suitability and environmental aesthetics: Conundrum, canard and curiosity. *Landscape and Urban Planning, 32*, 227–244.

Pregio, C., Breuste, J.-H., & Rojas, J. (2008). Perception and value of nature in urban landscapes: A comparative analysis of cities in Germany, Chile and Spain. *Landscape Online, 7*, 1–22.

Roovers, P., Hermy, M., & Gulinck, H. (2002). Visitor profile, perceptions in forests from a gradient of increasing urbanisation in Central Belgium. *Landscape and Urban Planning, 59*, 129–145.

Rydberg, D., & Falck, J. (2000). Urban forestry in Sweden from a silvicultural perspective: A review. *Landscape and Urban Planning, 47*, 1–18.

Spronken-Smith, R. A., & Oke, T. R. (1998). The thermal regime of urban parks in two cities with different summer climates. *International Journal of Remote Sensing, 19*, 2085–2104.

Standish, R. J., Hobbs, R. J., & Miller, J. R. (2013). Improving city life: Option for ecology restoration in urban landscapes and how these might influence interactions between people and nature. *Landscape Ecology, 28*, 1213–1221.

Steel, B., List, P., & Schindler, B. (1994). Conflicting values about federal forest: A comparison of national and Oregon public. *Society and Natural Resources, 7*, 137–153.

Stewart, W. P., Liebert, D., & Larkin, K. W. (2004). Community identities as visions for landscape change. *Landscape and Urban Planning, 69*, 315–334.

Tahvanainen, L., Tyrväinen, L., Ihalainen, M., Vuorela, N., & Kolehmainen, O. (2001). Forest management and public perception—Visual version verbal information. *Landscape and Urban Planning, 53*, 53–70.

Tzoulas, K., Korpela, K., Venn, S., Pelkonen, Y., Kazmierczak, A., Niemelä, J., & James, P. (2007). Promoting ecosystem and human health in urban areas using green infrastructure: A literature review. *Landscape and Urban Planning, 81*, 167–178.

Ulrich, R. S. (1984). View through a window may influence recovery from surgery. *Science, 224*, 420–421.

Ulrich, R. S. (1992). How design impacts wellness. *The Healthcare Forum Journal, 9*, 20–25.

Van den Berg, A. E., & Koole, S. (2006). New wilderness in The Netherlands: An investigation of visual preferences for nature development landscapes. *Landscape and Urban Planning, 78*, 362–372.

Yang, D., Luo, T., Lin, T., Qui, Q., & Luo, Y. (2014). Combining aesthetic with ecological values for landscape sustainability. *PLoS One, 9*(7), 1–7.

Yu, K. (1995). Cultural variations in landscape preference. Comparisons among Chinese subgroups and Western design experts. *Landscape and Urban Planning, 32*, 107–126.

Zhang, B. (2013). The aesthetic attributes of green infrastructure—A study of the perceptions of beauty, ecological significance, and naturalness for a stormwater treatment area by three college populations with different educational backgrounds. A dissertation presented to the graduate school of the University of Florida in partial fulfillment of the requirements for the degree of doctor of philosophy. University of Florida.

Zhang, Y., Zheng, B., & Laband, D. (2010). *Assessing preferences for and attitudes towards urban forests*. Auburn University.

Agriculture and Climate Change in Pakistan: A Review Study of Impacts on Wheat and Rice Production by 2050

Asad Aziz and Muhammad Mushahid Anwar

Abstract

Agriculture or agriculture industry is considered one of the most sensitive to climate sector of an economy. The previous research leads to a systematic model for intercomparisons between climate and its impacts on agriculture and agricultural production. However, varied results are obtained due to the nature of input data, model approaches, and different research scenarios. These scenarios will change the equilibrium of decision-making support system and economic modeling to investigate the climate change impacts on agriculture. In Pakistan's Real Gross Domestic Product, by "2050" climate change made a huge loss on the production of rice and wheat crops by "19.5 dollars," with also decreased domestic private consumption. This impact not only declines the crops' production but also has also some indirect effects on the industrial, business, and economic sectors, along with the agriculture sector of the country. These issues raise a huge number of challenges for the poor, particularly in urban areas or urban households. However, it is recommended that the government must implement a solid strategy related to agriculture that can enhance its capability to properly respond to climate change and minimize the challenges as much as possible.

Keywords

Agriculture · Agriculture industry · Agriculture production · Climate · Crops · Collaborative research economic modeling · Impacts · Pakistan · Production

Climate Change and Integrated Approaches

Scientists and agricultural specialists conducted integrated and collaborative research to investigate the possible effects of climate change (Mahrous, 2019). They used global climate models to examine the weather/climatic factors are interacting with utilizing diverse, biological, chemical, and physical factors, and forecast their reactions to growing amounts of emissions in the air and contribute to greenhouse gas (Illangasingha et al., 2023) (Wang et al., 2023). These parameters also take into account other socioeconomic scenarios, such as population growth, power consumption, industrial expansion, and income, in predicting the future climate of the earth (Liu et al., 2021) and (Ahmed et al., 2023). Along the side, these global climate estimates were for-

A. Aziz (✉) · M. M. Anwar
Department of Geography, University of Gujrat, Gujrat, Punjab, Pakistan
e-mail: asadaziz@uog.edu.pk; mushahid.anwar@uog.edu.pk

M. M. Anwar, A. Aziz (eds.), *Sustainable Landscape for Urban Resilience*, Sustainable Development Goals Series, https://doi.org/10.1007/978-3-031-99836-2_13

merly used to develop crop models by biophysical scientists to generate biological methods of crop development as well as productivity (Qu et al., 2019) (Pixley et al., 2023). They also evaluate the climate change and its impacts on agriculture output as well as on human health. But on the other hand, these models eliminate the requirement for expensive and prolong field surveys. These models eliminate the need for time-consuming and costly field surveys and experiments for the investigation of the impacts of climatic variables on cultivation in the agroindustry.

The possible influence of meteorological conditions on agricultural yield may be predicted using biophysical models to be used as a contribution in several limited and wide-ranging equilibrium on large-scale economic models to study economic data . Climate change responses by various socioeconomic forces of society as shown in Fig. 1.

The climate change and its impacts take long-term consequences for agricultural communities across the world (Spies, 2018). At the end of the twenty-first century, the average annual temperature in South Asia could rise with change of (3.3 °C) as reported by the Intergovernmental Panel on Climate Change (IPCC-5th) annual report (AR5) (Abeysingha et al., 2016; Almazroui et al., 2020).

A substantial quantity of research illustrates the serious issue of South Asian economies' agriculture sectors as a result of extreme weather conditions (Abeysekara et al., 2024; Colelli & Mistry, 2022). This industry will struggle to provide food security to the region's growing population (Anwar et al., 2023). Although more than 60% of total population is engaged in this sector, however, due to change of climate that affect and lose, the ratio of agricultural production is a serious concern in this region (Ahmed et al., 2018; Elahi et al., 2022).

In spite of contributing that is just 0.8% of greenhouse gases (GHG) in global atmosphere (Smoot & Baxter, 2003; Yue & Gao, 2018), the country has a vulnerable condition to climate change and ranks 135th in relation to the other countries (Nunes, 2023; Ruane et al., 2014). Regarding the geographical location and climatic variations, Pakistan was ranked 7th in the Global Climate Risk Index in 2017 (Aziz & Anwar, 2024). Pakistan is situated in a geographical location where the average temperature increases are predicted to be more compared to the world's average (Baocheng et al., 2024). Glaciers, the feeding source of rivers, melt dramatically, and the major area is semi-arid or dry (Sun et al., 2017). Greater than 40% of the country's population is involved in agriculture (Ali et al., 2017). The changing pattern of monsoon rains, droughts, and major floods enhances the risk to exposure in the region (Khan et al., 2020). The aggregated effects of all of these climate deviations place the country's food, water, and energy security in threat.

Despite growing research on climate change-initiated declines in production in the country (Carter et al., 2018), the degree and nature of their economic consequences still remain unknown (Abbass et al., 2022) (Habib-ur-Rahman et al., 2022). Because most of the previous studies count, only major climate change variables are included, including precipitation and temperature on total crops yield, which is

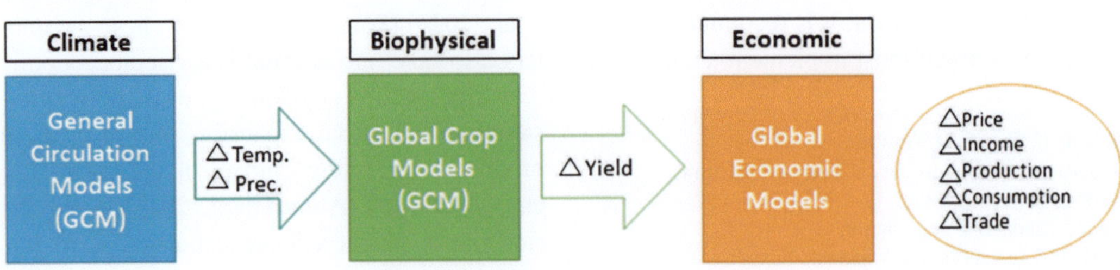

Fig. 1 Climate and global parameters. Adapted from Khan et al. (2020)

partial equilibrium analysis on climatic change or its impacts. Agricultural production is utilized as raw material both directly and indirectly. Any change in crop output would have an impact on the economy as a whole. As a result, the findings of previous analysis studies are unable to provide policymaking tools with comprehensive recommendations.

Methods

The study is conducted by considering the data set of two crops such as wheat and rice. The Pakistan Agricultural Research Council provided scientific data on the production of various crops, as well as data of temperature and precipitation (PARC) (Ali, 2018). The data for the selection of the review was provided by the Pakistan Meteorological and Pakistan Bureau of Statistics. The economic examination of impacts of climate change on production begins with the introduction of a set of parameters associated with climate change circumstances and crop production inputs data that can be used in the world economic model (Fróna et al., 2021). Using biophysical models, this combined modeling background connects universal climate models to the local economic model (Babar et al., 2021). Wheat and rice cultivation has been divided into three phases, with each stage having a distinct precipitation and optimal temperature. The data set on district-level agricultural productivity were obtained from the Ministry of Agriculture's statistical year book, while the statistics on precipitation with temperature were obtained from the metrology department (Abid et al., 2015). The study uses fixed effect model (FEM) for the analysis of the above data through regression model in which the production of both products was taken as a dependent variable while different stages on temporal basis temperature and precipitations were taken as independent variables.

The following equation clears the formulation of this case study (Adapted from Khan et al., 2020):

$$\text{Crops}\left(\text{W and R}\right) = f\,(\text{TST}^{\text{W}}, \text{TST}^{\text{R}}, \text{FST}^{\text{W}}, \text{FST}^{\text{R}}, \text{FP}^{\text{W}}, \text{FP}^{\text{R}}, \text{SSP}^{\text{W}}, \text{SSP}^{\text{R}}, \text{TSP}^{\text{W}}, \text{TSP}^{\text{R}}, \text{FSP}^{\text{W}}, \text{FSP}^{\text{R}}, \text{FT}^{\text{W}}, \text{FT}^{\text{R}}, \text{SST}^{\text{W}}, \text{SST}^{\text{R}})$$

$$(\text{Crops})_{it} = \alpha_i + \beta_1 (\text{FT})_{it} + \beta_2 (\text{FT}_2)_{it} \text{------} + \beta_n (\text{Tem,Pre})_{it} + V_{it} \mathbf{V}_{it} = \mu_i + \Sigma \mathbf{W}_{it}, \quad (\mathbf{t} = \mathbf{1}, \mathbf{2} \ldots \mathbf{T} : \mathbf{i} = \mathbf{1}, \mathbf{2} \ldots \mathbf{N})$$

Vit composite error term, μi i unobservable different country specific effects, $\sum$wit disturbances.

Results

Climate change is significant because of the wide-ranging socioeconomic consequences. Economist, researchers, and policymakers try to attempt to examine the potential consequences of climate change in the country. Consequences show that in Pakistan agriculture is a critical component of an economy, which provides raw materials to the industries and jobs to population of the country. The considered climatic change variables, including temperature and precipitation and temporal change in them, are estimated in the map in Fig. 2.

Figure 3 is used to show the economic impacts on the country's economy on total GDP, population, total import, and export in a year:

Figure 4 shows the impacts of climate-related yield losses in rice and wheat on key sector output and consumption. With a stable and unchanging environment, wheat and rice output will almost increase by 2050. The production of agricultural goods is predicted to stagnate with the passage of time under harsh climatic scenarios. Rice and wheat output are predicted to fall by 15–20%, respectively, by 2050.

Decrease in the production of a goods raises the rate/cost and reduces consumption. The buying power is directly linked with GDP, which estimates the economic power of a country at a given time. The immediate result of climate-

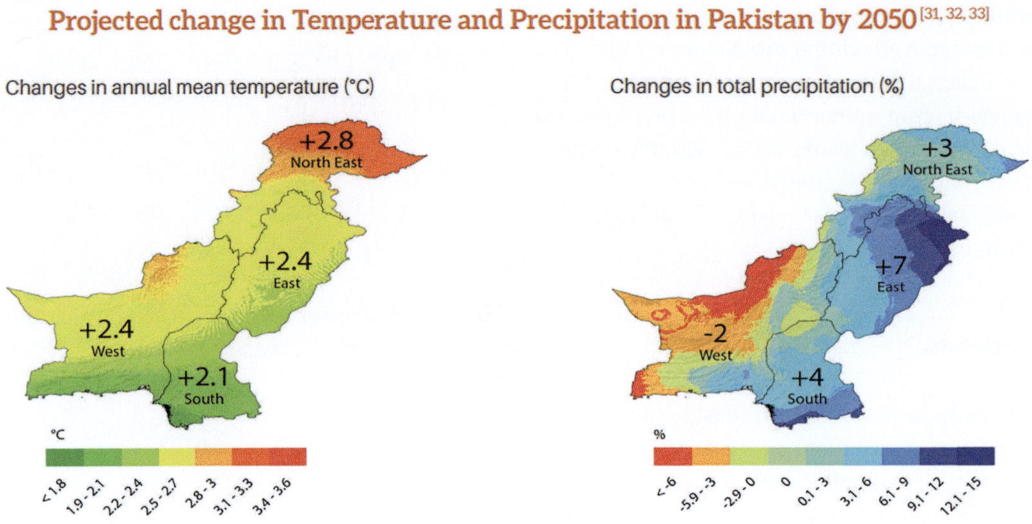

Fig. 2 Projected map showing climate variables. Adapted from Chaudhry (2017)

Economic variables	Baseline 2050 (Million $)	Simulated Shocks % Change (Monetary Change)
Real GDP	527,804	−3.7 (−19,528)
Population	319.37	-
Real Exports	74,770	−2.3 (1719.8)
Real Imports	105,927	1.54 (1631.2)

Fig. 3 Economic impacts of climate change in Pakistan. (Source: Khan et al., 2020)

Fig. 4 Impacts of climate change on domestic price. (Source Khan et al., 2020)

Crops	Sectoral Production	
	Baseline 2050	Simulated Shocks
	Million ($)	% Change
Grain Crops	40,727	−0.01
Paddy Rice	11,027	−20
Wheat	15,399	−14.6

induced agricultural yield loss is an increase in commodities prices.

To recognize the significant damage in the country's GDP as a result of agricultural production declines in specific climatic scenarios, a valuation of the individual factors that control actual GDP by using economic parameters such as employment wages, profits, capital interests, taxes, and land rent is done. It has been noted that the decrease in output generates a deficiency in the market, which causes rise in prices in any country.

Discussion

The research demonstrates the adverse significance of climate change in crops production on total economy by 2050. Although agriculture is the most influential sector of economy, contributed in GDP. For any decrease in agro-production, the climate change impacts not just in this sector, but its consequences may spread to agricultural-associated and its connected businesses, such as services and manufacturing. The study identifies some serious policy consequences of a wider angle on climate change. To help small- to medium-class agriculturalists or farmers, even agricultural employees, the government must implement long-term climate-change initiatives.

Given the possibility of future climate or associated disasters, particularly floods, the government must invest water managing technology and irrigation system. The government must support agricultural farmers with farm advice services, weather forecasts, and marketing information. It is proposed that the government have a solid agricultural strategy that can influence its capability to successfully adjust to change of climate, which is only required for more agricultural output and agricultural sustainability.

Conclusion

This research, like every other type of economic research, has limitations. It investigates the upcoming response of an agricultural economic modeling to multiple climatic change and its consequences by 2050, focusing solely on economic element and their assessment. It does not go into depth into the biophysical elements of climate change impacts, such as assessing the real cost of agricultural production loss. As a result, crop modeling research, in association with the production of agriculture, is very necessary to investigate real-time analysis for crops yield estimations. Besides, an arithmetical valuation of the effects and potential response to change of climate may need a considerable extended modeling and framework. Regardless of the limits mentioned above, the findings of the study show that substantial policy planning and its application of adaption techniques are needed in the future or near future to assist and mitigate the detrimental significance of climate change on agriculture.

References

Abbass, K., Qasim, M. Z., Song, H., Murshed, M., Mahmood, H., & Younis, I. (2022). A review of the global climate change impacts, adaptation, and sustainable mitigation measures. *Environmental Science and Pollution Research, 29*(28), 42539–42559. https://doi.org/10.1007/s11356-022-19718-6

Abeysekara, W. C. S. M., Siriwardana, M., & Meng, S. (2024). Economic consequences of climate change impacts on South Asian agriculture: A computable general equilibrium analysis. *Australian Journal of Agricultural and Resource Economics, 68*(1), 77–100. https://doi.org/10.1111/1467-8489.12541

Abeysingha, N. S., Singh, M., Islam, A., & Sehgal, V. K. (2016). Climate change impacts on irrigated rice and wheat production in Gomti River basin of India: a case study. *Springerplus, 5*(1), 1250. https://doi.org/10.1186/s40064-016-2905-y

Abid, M., Scheffran, J., Schneider, U. A., & Ashfaq, M. (2015). Farmers' perceptions of and adaptation strategies to climate change and their determinants: the case of Punjab province, Pakistan. *Earth System Dynamics, 6*(1), 225–243. https://doi.org/10.5194/esd-6-225-2015

Ahmed, M., Huan, W., Ali, N., Shafi, A., Ehsan, M., Abdelrahman, K., Khan, A. A., Abbasi, S. S., & Fnais, M. S. (2023). The Effect of energy consumption, income, and population growth on CO2 emissions: Evidence from NARDL and machine learning models. *Sustainability, 15*, 15.

Ahmed, M., Ijaz, W., & Ahmad, S. (2018). Adapting and evaluating APSIM-SoilP-Wheat model for response to phosphorus under rainfed conditions of Pakistan. *Journal of Plant Nutrition, 41*(16), 2069–2084. https://doi.org/10.1080/01904167.2018.1485933

Ali, G. (2018). Climate change and associated spatial heterogeneity of Pakistan: Empirical evidence using multidisciplinary approach. *Science of the Total Environment, 634*, 95–108. https://doi.org/10.1016/j.scitotenv.2018.03.170

Ali, S., Liu, Y., Ishaq, M., Shah, T., Abdullah, Ilyas, A., & Din, I. U. (2017). Climate change and its impact on the yield of major food crops: Evidence from Pakistan. *Food, 6*(6), 39. https://doi.org/10.3390/foods6060039

Almazroui, M., Saeed, S., Saeed, F., Islam, M. N., & Ismail, M. (2020). Projections of precipitation and temperature over the south Asian countries in CMIP6. *Earth Systems and Environment, 4*(2), 297–320. https://doi.org/10.1007/s41748-020-00157-7

Anwar, M. M., Breuste, J. H., Ahmad, A., Aziz, A., & Aldosari, A. A. (2023). Quantifying the impacts of urbanization on urban agriculture and food security in the megacity Lahore, Pakistan. *Sustainability, 15*, 16.

Aziz, A., & Anwar, M. M. (2024). Assessing the impacts of urban expansion on sustainability; empirical evidence from Islamabad, Pakistan. *GeoJournal, 89*(2), 82. https://doi.org/10.1007/s10708-024-11087-6

Babar, M. S., Tazyeen, S., Khan, H., Tsagkaris, C., Essar, M. Y., & Ahmad, S. (2021). Impact of climate change on health in Karachi, Pakistan. *The Journal of Climate Change and Health, 2*, 100013. https://doi.org/10.1016/j.joclim.2021.100013

Baocheng, H., Jamil, A., Bellaoulah, M., Mukhtar, A., & Clauvis, N. K. (2024). Impact of climate change on water scarcity in Pakistan. Implications for water management and policy. *Journal of Water and Climate Change, 15*(8), 3602–3623. https://doi.org/10.2166/wcc.2024.710

Carter, C., Cui, X., Ghanem, D., & Mérel, P. (2018). Identifying the economic impacts of climate change on agriculture. *Annual Review of Resource Economics, 10*, 22938. https://doi.org/10.1146/annurev-resource-100517-022938

Chaudhry, Q. U. Z. (2017). *Climate change*. Asian Development Bank. https://www.adb.org/publications/climate-change-profile-pakistan

Colelli, F. P., & Mistry, M. N. (2022). Income-dependent expansion of electricity demand for climate change adaptation in Brazil. *Energy and Climate Change, 3*, 100071. https://doi.org/10.1016/j.egycc.2022.100071

Elahi, E., Khalid, Z., Tauni, M. Z., Zhang, H., & Lirong, X. (2022). Extreme weather events risk to crop-production and the adaptation of innovative management strategies to mitigate the risk: A retrospective survey of rural Punjab. *Technovation, 117*, 102255. https://doi.org/10.1016/j.technovation.2021.102255

Fróna, D., Szenderák, J., & Harangi-Rákos, M. (2021). Economic effects of climate change on global agricultural production. *Nature Conservation, 44*, 117–139. https://doi.org/10.3897/natureconservation.44.64296

Habib-Ur-Rahman, M., Ahmad, A., Raza, A., Hasnain, M. U., Alharby, H. F., Alzahrani, Y. M., Bamagoos, A. A., Hakeem, K. R., Ahmad, S., Nasim, W., Ali, S., Mansour, F., & El Sabagh, A. (2022). Impact of climate change on agricultural production; Issues, challenges, and opportunities in Asia. *Frontiers in Plant Science, 13*, 925548. https://www.frontiersin.org/journals/plant-science/articles/10.3389/fpls.2022.925548

Illangasingha, S., Koike, T., Rasmy, M., Tamakawa, K., Matsuki, H., & Selvarajah, H. (2023). A holistic approach for using global climate model (GCM) outputs in decision making. *Journal of Hydrology, 626*, 130213. https://doi.org/10.1016/j.jhydrol.2023.130213

Khan, M. A., Tahir, A., Khurshid, N., Husnain, M. I., Ahmed, M., & Boughanmi, H. (2020). Economic effects of climate change-induced loss of agricultural production by 2050: A case study of Pakistan. *Sustainability, 12*(3), 1216. https://doi.org/10.3390/su12031216

Liu, S., Zeng, A., Lau, K., Ren, C., Chan, P.-W., & Ng, E. (2021). Predicting long-term monthly electricity demand under future climatic and socioeconomic changes using data-driven methods: A case study of Hong Kong. *Sustainable Cities and Society, 70*, 102936. https://doi.org/10.1016/j.scs.2021.102936

Mahrous, W. (2019). Climate change and food security in EAC region: A panel data analysis. *Review of Economics and Political Science, 4*(4), 270–284. https://doi.org/10.1108/REPS-12-2018-0039

Nunes, L. J. R. (2023). The rising threat of atmospheric CO_2: A review on the causes, impacts, and mitigation strategies. *Environments, 10*(4), 66.

Pixley, K. V., Cairns, J. E., Lopez-Ridaura, S., Ojiewo, C. O., Dawud, M. A., Drabo, I., Mindaye, T., Nebie, B., Asea, G., Das, B., Daudi, H., Desmae, H., Batieno, B. J., Boukar, O., Mukankusi, C. T. M., Nkalubo, S. T., Hearne, S. J., Dhugga, K. S., Gandhi, H., & Zepeda-Villarreal, E. A. (2023). Redesigning crop varieties to win the race between climate change and food security. *Molecular Plant, 16*(10), 1590–1611. https://doi.org/10.1016/j.molp.2023.09.003

Qu, C.-H., Li, X.-X., Hui, J., & Liu, Q. (2019). ScienceDirect the impacts of climate change on wheat yield in the Huang-Huai-Hai plain using DSSAT-CERES-wheat model under different climate scenarios. *Journal of Integrative Agriculture, 18*, 60345–60352. https://doi.org/10.1016/S2095-3119(19)62585-2

Ruane, A. C., Winter, J. M., McDermid, S. P., & Hudson, N. I. (2014). AgMIP climate data and scenarios for integrated assessment. In *Handbook of climate change and agroecosystems* (Vol. 3, pp. 45–78). Imperial College Press. https://doi.org/10.1142/9781783265640_0003

Smoot, L. D., & Baxter, L. L. (2003). Fossil fuel power stations—Coal utilization. In R. A. Meyers (Ed.), *Encyclopedia of physical science and technology* (3rd ed., pp. 121–144). Academic Press. https://doi.org/10.1016/B0-12-227410-5/00257-X

Spies, M. (2018). Changing food systems and their resilience in the Karakoram mountains of Northern Pakistan a case study of nagar. *Mountain Research and Development, 38*(4), 299–309.

Sun, X., Wang, K., Kang, S., Guo, J., Zhang, G., Huang, J., Cong, Z., Sun, S., & Zhang, Q. (2017). The role of melting alpine glaciers in mercury export and transport: An intensive sampling campaign in the Qugaqie Basin, inland Tibetan Plateau. *Environmental Pollution, 220*, 936–945. https://doi.org/10.1016/j.envpol.2016.10.079

Wang, C., Song, J., Shi, D., Reyna, J. L., Horsey, H., Feron, S., Zhou, Y., Ouyang, Z., Li, Y., & Jackson, R. B. (2023). Impacts of climate change, population growth, and power sector decarbonization on urban building energy use. *Nature Communications, 14*(1), 6434. https://doi.org/10.1038/s41467-023-41458-5

Yue, X.-L., & Gao, Q.-X. (2018). Contributions of natural systems and human activity to greenhouse gas emissions. *Advances in Climate Change Research, 9*(4), 243–252. https://doi.org/10.1016/j.accre.2018.12.003

Urban Resilience, Development, and Sustainability: A Geographical Evaluation, from Research to Action (Concept Notes)

Asad Aziz and Muhammad Mushahid Anwar

Abstract

Urban resilience is the capability of urban world to act efficiently in such a way that their residents and working population, particularly the vulnerable people, undergo the shocks and stresses that they encounter in their normal lives. This study makes an attempt to evaluate the impact of climate change on low-income population, equity, health, and well-being in poor urban. Climate change as global social, economical, biological, and natural challenge and threat spatially has two major impacts; first, it affects the poor and vulnerable people; and second, it has long-standing effects for sustainable development. It has been investigated that poor, older age, low economic status, and female are more at risk and vulnerable to climate change and its impacts in urban areas in the globe. This impact is more dominated in low-income residencies and developing nations, where the population lacks basic facilities, low-nutrition food, and poor social, environmental, and natural services. As a result of this crucial issue, this impact becomes gradually deceptive; it is a matter of fact that those who are at most risk will get most affected socially as well as economically. This impact generates a newly emerging term "climatic gentrification," which is getting more importance around the world among policymakers. The impact of the term gentrification is a crucial issue not only for urban resilience but also create long-term challenges for urban sustainability. The study result will provide fundamental results after a careful and critical investigation to create awareness among policymakers and urban governance to make attempts to minimize this issue to the possible level. Finally, the outcomes of the study will help create an environment for social justice all over the world for urban resilience.

Keywords

Climate gentrification · Development · Sustainability · Urban · Geography · Urban · Resilience

Introduction

To respond to disasters, stresses, and uncertainties, an increasing number of cities are integrating resilience into their plans and policies that is an indirect impact of climate change for mitigation. This impact brings the government and every individual to begin the projected efforts to face this issue for their existence. The literature

A. Aziz (✉) · M. M. Anwar
Department of Geography, University of Gujrat, Gujrat, Punjab, Pakistan
e-mail: asadaziz@uog.edu.pk; mushahid.anwar@uog.edu.pk

M. M. Anwar, A. Aziz (eds.), *Sustainable Landscape for Urban Resilience*, Sustainable Development Goals Series, https://doi.org/10.1007/978-3-031-99836-2_14

found that the whiter and wealthier people moved to less exposed areas with little exposure, whereas the people occupied by low income are greatly exposed to disaster (Rhode, 2021). This means a high-income group of people will have less risk of exposure because of high socio-economic status. In other words, it is an example of social injustice. The climate change, sea-level rise, melting of glaciers, and other climatic issues, including global warming, affect the whole world in different phases with different impacts (Haasnoot et al., 2021). These natural disasters indirectly generate social issue, which challenged the well-being of humanity, particularly in the urban world, where the urban world in developing countries is a complicated network of services and environmental values. However, this effect has another face in the Global South with different scenarios (Gould & Lewis, 2021)

The countries need an immediate action to alleviate these issues in most vulnerable areas by implementing different environmental protection strategies, resilience planning strategies, and climate adaptation plans (Institute, 2019) to boost the urban resilience level. However, as the magnitude of these climatic impact is of a large scale, these actions still result in inequitable outcomes everywhere, where adopted by some countries. The process and outcomes of policies regarding the climatic gentrification are highly dependent upon the context and ultimately increase the disparities among the communities (University, 2021). However, without social justice, considering the policies and strategies for urban resilience is vulnerable for low-income communities. The intervention in these population and communities also aggravates inequities economically as well as temporally.

Cities in less developed nations are more at risk and estimated to lose more as people have low socio-economic profile and urban infrastructure is less developed to resist the natural disaster. It has been noted that in developing countries even with more population with low socio-economic status get affected the most in economic loss, housing displacement, and more pandemic diseases due to malnutrition and inequity in the provision of food. This type of social injustice has been noted in many of the countries in the entire world. This social injustice creates many challenges of peace in the society, sustainable development, and urban resilience (Martyr-Koller et al., 2021). This social injustice introduced a new term when it is associated with climate change and its impacts, which is commonly known as "climatic gentrification."

The term climatic gentrification was introduced at Harvard University by "Anurag Gumber, Thomas Hill and Jesse Keenan" (Keenan et al., 2018). This term can be described in multidomain according to the nature of even or phenomena happening in real time. It has been seen the current rate of urbanization is not only in developing world; this phenomenon is going to be increased globally and disturb the natural cycle of resources and putting pressure on social, environmental, and political values.

Research Methodology

The methodology used in the study has two fundamental frames, including qualitative and quantitative nature of data for adoption and mitigation plan to respond to a disaster. This includes the geographic information system (GIS) for spatial identification of the problem of urban areas related to the current issue under examination with respect to the study area. The related issue proposed for the analysis is in urban green spaces and equity related to health, social justice, and sustainability in socio-economic profile for a geographical analysis. The spatial investigation involves the examination of area under study and make deduction; after careful investigation of these deductions, the methods shall be applied to minimize the issue for urban resilience and development.

This research is classified as descriptive in terms of its aims into rational and irrational elements as indicated in Fig. 1. The research is bibliographic in terms of technical techniques, and it is documental in terms of data collection. This

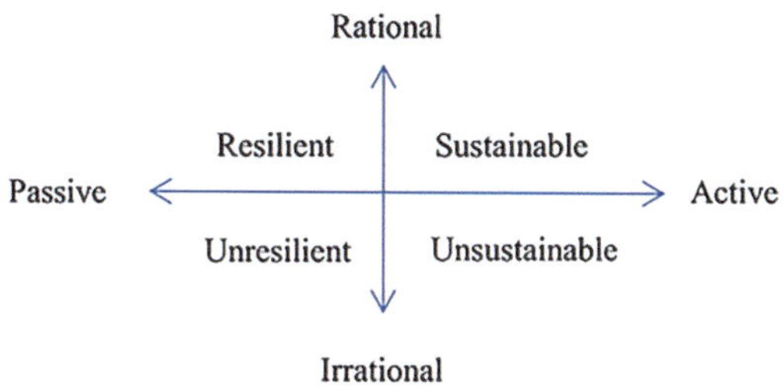

Fig. 1 Urban resilience and regional approach

study takes a quantitative approach since it employs the statistical method.

Discussion

Climate change as a response of the natural disaster is affecting every individual in all aspects of life, including environmental, economic, and social. It has been noted that people with low-income status affect the most and more vulnerable than other. On the other hand, climate change also affects them socially in the form of change in rain fall pattern that indirectly decreased the agricultural production ratio, and flash flooding that damages the poor house settlements in urban to rural areas. This impact drives a term known as climatic gentrification and induced the social injustice for urban resilience. This study is formulated to analyze the impacts of climate change/gentrification on urban resilience by taking environmental, economic, and social domains into account and also evaluate its relation with social justice for urban development. The previous studies provide forceful evidences in this regard and open the door to investigate further how to overcome these challenges, particularly in the developing world.

Conclusions

The study provides evidence and well-investigated outcomes to bring attention to urban managers and government regarding the most vulnerable issues. These outcomes of the study will be helpful for future development planning regarding disaster management, social development, and planning strategies to making environmental laws. The assessment of these social attributes is a formal way to predict the disaster outcomes to overcome its impacts. If comparative studies will be done on such crucial issues on affected areas, fruitful results could be obtained in the form of disaster management and urban development plans for residential, formal, and informal settlements with social justices in urban resilience scansions. GIS tools and satellite imagery will provide logical-based evidence in terms of spatially weighted and statistically measured evidence and related data. The results of this study will be the solid evidence based on actual values and show significant effects, which support the allocation plans for settlements, risk control to minimize the economic loss, and exposure and city development to make it easy to achieve the United Nations sustainable development goals.

References

Gould, K. A., & Lewis, T. L. (2021). Resilience gentrification: Environmental privilege in an age of coastal climate disasters. *Frontiers in Sustainable Cities, 3*, 85.

Haasnoot, M., Winter, G., Brown, S., Dawson, R. J., Ward, P. J., & Eilander, D. (2021). Long-term sea-level rise necessitates a commitment to adaptation: A first order assessment. *Climate Risk Management, 34*, 100355. https://doi.org/10.1016/j.crm.2021.100355

Institute, E. L. (2019). *Climate gentrification and resilience planning: What is at stake for at-risk communities?* Environmental Law Institute. Retrieved November 15, 2021, from https://www.eli.org/vibrant-environment-blog/climate-gentrification-and-resilience-planning-what-stake-risk-communities

Keenan, J., Hill, T., & Gumber, A. (2018). Climate gentrification: From theory to empiricism in Miami-Dade County. *Florida. Environmental Research Letters, 13*, 054001. https://doi.org/10.1088/1748-9326/aabb32

Martyr-Koller, R., Thomas, A., Schleussner, C.-F., Nauels, A., & Lissner, T. (2021). Loss and damage implications of sea-level rise on Small Island Developing States. *Current Opinion in Environmental Sustainability, 50*, 245–259. https://doi.org/10.1016/j.cosust.2021.05.001

Rhode, R. (2021). Sea level rise threatens Miami's future. Here's how the Army Corps can help keep Florida, Florida. Sea level rise threatens Miami's future. Retrieved November 23, 2021, from https://blogs.edf.org/growingreturns/2021/07/23/sea-level-rise-threatens-miamis-future/

University, N. (2021). *Climate gentrification: Why we need to consider social justice in climate change planning*. Resilience News. Retrieved May 11, 2021, from https://globalresilience.northeastern.edu/climate-gentrification-why-we-need-to-consider-social-justice-in-climate-change-planning/

Conclusion: Toward a Sustainable Landscape for Urban Resilience

Muhammad Mushahid Anwar

Abstract

As this book draws to a close, it is important to revisit the foundational principles that have guided its structure: the urgent need for resilient cities, the role of sustainable landscapes in urban adaptation, and the importance of integrating ecological, technological, and social systems into planning frameworks. The case studies of Islamabad and Bucharest have demonstrated that, despite differing socio-political contexts, urban areas around the world are experiencing converging pressures—from climate change and biodiversity loss to spatial inequality and governance fragmentation.

The concept of urban resilience, while widely adopted, has often been applied in narrow technical or infrastructural terms. However, as outlined throughout this volume, resilience is a multidimensional construct that must encompass ecological regeneration, socio-economic equity, participatory governance, and cultural recognition (Meerow et al., Landsc Urban Plan 147:38–49, 2016). When viewed through the lens of landscape planning, resilience becomes not just a property of infrastructure, but a quality of the entire urban ecosystem—including its people, places, and the relationships that bind them.

Sustainable landscapes, then, are not ancillary to urban development—they are foundational. Landscapes regulate temperature, filter pollutants, manage water, and support biodiversity. But they also shape social behavior, enable recreation, provide cultural identity, and serve as democratic public spaces. Their value lies not only in their environmental function but also in their symbolic and spatial meaning within the city (Forman, Urban ecology: science of cities. Cambridge University Press, Cambridge, 2014; Agyeman et al., Just sustainabilities: development in an unequal world. MIT Press, Cambridge, 2003).

The cities studied in this book, Islamabad and Bucharest, offer distinct yet complementary lessons. Islamabad, despite its origins as a master-planned city, has witnessed rapid peri-urban expansion, weakening governance, and ecological degradation due to land market dynamics and limited enforcement of zoning regulations (Anwar and Aziz, J Urban Ecol Sustain (In press), 2025). Bucharest, shaped by its socialist legacy and market-driven transformation, faces institutional fragmentation and social inequity in the distribution of green infrastructure (Gavrilidis et al., Urban For Urban Green 78:127778, 2022).

In both contexts, landscape-based interventions such as urban forests, green corridors,

M. M. Anwar (✉)
Department of Geography, University of Gujrat,
Gujrat, Punjab, Pakistan
e-mail: mushahid.anwar@uog.edu.pk

M. M. Anwar, A. Aziz (eds.), *Sustainable Landscape for Urban Resilience*, Sustainable Development Goals Series, https://doi.org/10.1007/978-3-031-99836-2_15

small public urban green spaces (SPUGS), and nature-based solutions have shown great promise. Yet they also face persistent challenges: underfunding, poor maintenance, lack of coordination, and limited citizen participation. These obstacles suggest that technical solutions alone are insufficient. What is required is a reframing of planning paradigms one that places landscapes not at the margins but at the core of resilient urban development.

This final chapter consolidates the themes of the previous sections and presents a forward-looking roadmap for policy, practice, and research. It argues for a bold rethinking of how cities are built, how decisions are made, and how nature and society can be reintegrated in an era of rapid transformation.

Keywords

Analysis · Urban · Planning · Further · Islamabad

Lessons from Islamabad: Gaps, Potentials, and Future Directions

Islamabad, conceived in the 1960s as a symbol of modernist planning in South Asia, was envisioned as a city in harmony with nature. Surrounded by the Margalla Hills, intersected by natural streams, and structured with green belts, Islamabad initially represented an ideal blend of ecology and design. However, over time, this vision has eroded under the pressure of population growth, administrative inertia, and market-driven development. The case of Islamabad offers a valuable lens into the gaps and potentials that define landscape resilience in a rapidly urbanizing context.

One of the most significant findings from the Islamabad chapters is the unregulated expansion of built-up areas, particularly in Zones IV and V. GIS-based land-use studies revealed that between 2000 and 2020, vegetative and agricultural lands declined while commercial and residential zones expanded by over 130% (Anwar & Aziz, 2025). This transformation is not just spatial but systemic—driven by speculative real estate development, lack of enforcement of the city's master plan, and weak inter-agency coordination among planning institutions (Forman, 2014).

Furthermore, the case highlighted the fragmentation of ecological buffers, such as green belts and riparian corridors. These spaces were originally intended to mitigate heat, manage flood risks, and support biodiversity. Yet today, they are under threat from encroachment, illegal construction, and poor maintenance. This not only reduces the ecological resilience of the city but also compromises public health and safety, especially in lower-income and peri-urban communities.

Despite these challenges, Islamabad holds significant untapped potential. The city still retains large patches of forest, water bodies, and natural slopes that can be rehabilitated through ecological restoration. The existing zoning framework, if enforced, provides a solid foundation for reintegrating green infrastructure into urban form. Citizen-driven initiatives such as tree plantations, urban gardening, and environmental education campaigns show promise for creating a participatory culture around sustainability.

The analysis also underscored the importance of data-driven governance. Tools like NDVI and UAVs have allowed researchers and planners to monitor land degradation, heat islands, and vegetation loss with precision. These technologies can support real-time environmental monitoring and inform adaptive planning. However, their effectiveness depends on institutional will, technical capacity, and accessibility of geospatial data.

From a policy perspective, Islamabad requires a multilevel urban governance reform. This includes strengthening the mandate and coordination of the Capital Development Authority (CDA), aligning the Revised Master Plan with contemporary sustainability principles, and engaging citizens through inclusive decision-making platforms. Additionally, there is a need to revise building regulations to promote vertical expansion and minimize land sprawl.

In terms of future directions, integrating nature-based solutions (NBS) into planning at the sector level could provide ecological and social benefits. For example, restoring green corridors in flood-prone areas or incentivizing green roofs in commercial districts can enhance resilience without large-scale infrastructure costs. Furthermore, cross-sector collaboration—with academia, NGOs, and private developers—can foster innovation and accountability.

Islamabad's experience shows that even in cities with a strong planning legacy, sustainability is not guaranteed. The key lies in governance, public engagement, and reimagining landscapes as critical infrastructure—essential not only for resilience, but for the cultural and ecological identity of the city.

Lessons from Bucharest: Equity, Micro-Interventions, and Transformation

Bucharest, Romania's capital, offers a striking contrast to Islamabad in terms of historical development, socio-political context, and urban morphology. While Islamabad was conceived as a planned city from its inception, Bucharest evolved over centuries and underwent intense transformation during and after the socialist era. This legacy of centralized control, followed by a rapid neoliberal shift, has left the city with fragmented infrastructure, uneven development, and complex institutional arrangements. Despite these challenges, Bucharest serves as a compelling case study in the role of equity, micro-interventions, and adaptive governance in sustainable urban landscape planning.

One of the most revealing findings from the Bucharest chapters was the existence of stark spatial disparities in green space distribution. While the city contains major public parks such as Herăstrău and Carol Park, smaller neighborhood-level green spaces are inequitably distributed. Research showed that residents in lower-income districts like Ferentari and parts of Sector 5 often have less than 3 m^2 of green space per capita, compared to over 15 m^2 in wealthier central districts (Gavrilidis et al., 2022). This not only limits access to recreational and health benefits, but also deepens feelings of exclusion and marginalization.

Yet, the city also demonstrates the transformative power of Small Public Urban Green Spaces (SPUGS). These modest interventions—ranging from pocket parks and green alleys to repurposed vacant lots—have had disproportionate positive impacts. UAV and thermal imagery studies revealed that such spaces can reduce land surface temperatures by 2–3 °C during peak summer periods, improving microclimatic conditions for vulnerable populations. Moreover, these spaces foster local identity, intergenerational interaction, and informal social surveillance, contributing to community resilience.

Bucharest's experience also highlights the critical role of community engagement and grassroots innovation. In many cases, it was not municipal action but civic initiatives—volunteer greening programs, schoolyard gardens, and neighborhood tree-planting drives—that catalyzed ecological improvements. These bottom-up efforts often emerged in response to gaps left by a fragmented and sometimes indifferent institutional system. The city's decentralized district governance model creates both opportunities and limitations: while local administrations have the autonomy to innovate, coordination and standardization across sectors remain weak.

The case of Bucharest further illustrates how micro-level planning can deliver macro-level resilience benefits when embedded in a clear strategic vision. Strategic documents like the "Bucharest Green City Action Plan" and efforts to align with the EU Green Deal show potential, but require stronger policy backing, sustained funding, and a focus on long-term maintenance. Importantly, these efforts must incorporate social equity as a core principle, ensuring that benefits reach historically underserved populations.

Technology played a complementary role in the Bucharest study. High-resolution spatial data and UAV mapping enabled fine-grained environmental analysis and exposed gaps that would have been invisible in traditional urban audits.

However, institutional reluctance to adopt innovative tools remains a challenge. Bridging the gap between data and decision-making requires both technical training and policy mandates.

In conclusion, Bucharest teaches us that equity and resilience are inseparable. Addressing climate and ecological challenges in cities cannot succeed without a parallel focus on social inclusion. Micro-interventions—when integrated with strategic planning, community participation, and inter-agency coordination—can act as catalysts for systemic transformation. For post-socialist cities, this model offers a pathway to not only adapt to twenty-first-century urban pressures, but also to heal the spatial and social wounds of the past.

Comparative Reflections: Convergence, Divergence, and Synthesis

The comparative analysis of Islamabad and Bucharest, though grounded in distinct geographic, cultural, and political contexts, offers a unique opportunity to reflect on the broader challenges and opportunities in advancing sustainable urban landscapes. By juxtaposing a planned South Asian capital with a post-socialist Eastern European metropolis, this book has uncovered both converging patterns and contextual divergences in how cities are navigating sustainability, resilience, and equity in their urban transformations.

Convergence: Shared Challenges and Universal Patterns

Despite their differences, both cities exhibit similar systemic pressures: unplanned urban expansion, fragmented green infrastructure, rising vulnerability to climate change, and socio-spatial inequality. In both Islamabad and Bucharest, the degradation of ecological buffers—whether through unchecked real estate development in Islamabad or the privatization of communal spaces in Bucharest—has led to measurable environmental decline, including loss of biodiversity, increased heat stress, and deteriorating public health.

Institutionally, both cities suffer from governance fragmentation and weak enforcement. Islamabad's planning authorities, including the Capital Development Authority (CDA), often operate in silos, with overlapping jurisdictions and limited capacity for implementation. Bucharest, meanwhile, is governed by a multisectoral district model that decentralizes power but also produces inconsistencies and jurisdictional ambiguities. These conditions contribute to a policy environment where strategic planning exists on paper but struggles to manifest in practice.

Another point of convergence is the emergence of grassroots and community-driven initiatives. From Islamabad's volunteer-led tree plantation drives to Bucharest's citizen-maintained pocket parks, both cities are witnessing bottom-up efforts that compensate for institutional inertia. These initiatives highlight the critical role of civil society in landscape resilience and underscore the importance of participatory planning (Meerow et al. 2016).

Divergence: Institutional Histories and Planning Paradigms

While both cities share challenges, their divergence lies in the historical and structural roots of these challenges. Islamabad was born out of a post-colonial vision of technocratic planning and spatial order, designed with an emphasis on zoning, green belts, and centralized authority. Its decline in sustainability stems largely from institutional erosion, market pressures, and socio-political compromise on regulatory frameworks.

Bucharest's trajectory, by contrast, is shaped by socialist-era centralization, followed by an abrupt transition to capitalist development. The privatization of land and housing stock in the 1990s led to a patchwork of land ownership, making coordinated urban interventions more difficult. Additionally, the lack of participatory culture during the socialist period created a civic vacuum that persists today, despite growing awareness and activism.

In terms of technology adoption, Islamabad's use of NDVI and GIS tools has focused primarily on monitoring urban sprawl and vegetation loss. Bucharest has gone further in employing UAV-based thermal mapping, providing deeper insight into the microclimatic effects of urban greenery. However, Bucharest's integration of these tools into policymaking remains inconsistent, often confined to academic or pilot studies.

Synthesis: Toward a Shared Learning Agenda

Together, Islamabad and Bucharest highlight the need for a shared urban learning agenda that moves beyond isolated interventions. Sustainable landscapes must be embedded within a cross-sectoral, multi-scalar framework that includes institutional reform, civic engagement, ecological science, and technological innovation. Both cities offer valuable lessons: Islamabad shows the dangers of political capture of planning institutions; Bucharest illustrates the long-term impacts of fragmentation and the potential of small-scale interventions.

By recognizing both commonalities and contextual differences, urban resilience can be reframed not as a fixed model, but as a flexible, place-based approach. This synthesis provides the intellectual foundation for the next and final section, where key policy recommendations and future research directions are presented.

Policy Recommendations and Future Research

The insights derived from the comparative study of Islamabad and Bucharest underscore the urgent need to reframe urban policy through a lens of resilience, equity, and ecological intelligence. While both cities face unique governance and spatial challenges, the convergence of their urban pressures calls for a universal yet locally adaptable set of policy interventions that can guide future development. This final section presents actionable policy recommendations and identifies key areas for further academic and practical inquiry.

Policy Recommendations

1. Institutional Integration and Multi-Level Governance

Both Islamabad and Bucharest suffer from fragmented governance. A foundational recommendation is to enhance institutional integration across national, municipal, and district levels. This requires
   (a) Clarifying mandates between planning agencies and local authorities
   (b) Establishing inter-agency coordinating bodies for green infrastructure management
   (c) Embedding sustainability criteria into building codes and urban design guidelines

2. Mainstreaming Nature-Based Solutions (NBS)

Nature-based solutions should be institutionalized within city planning processes. This can be achieved through
   (a) Mandating green roofs, bioswales, and rain gardens in urban zoning plans
   (b) Providing tax incentives or subsidies for developers implementing NBS
   (c) Integrating NBS targets into national climate adaptation strategies (Kabisch et al., 2017)

3. Spatial Equity in Green Infrastructure

Governments must ensure the equitable distribution of green infrastructure, particularly in underserved neighborhoods. This includes
   (a) Implementing spatial justice audits using GIS and demographic overlays
   (b) Prioritizing SPUGS and urban reforestation in low-income districts (Gavrilidis et al., 2022)
   (c) Enforcing minimum green space standards per capita at the municipal level

4. Community Engagement and Participatory Planning

Public involvement should be formalized in landscape planning through

   (a) Participatory mapping exercises and citizen science platforms
   (b) Green infrastructure co-management models (e.g., park committees)
   (c) Integrating environmental education into local school curricula (Agyeman et al., 2003)

5. Data Transparency and Technological Capacity Building

Decision-makers need access to accurate, timely data. Cities should

   (a) Develop open-source urban ecological dashboards
   (b) Train planning staff in GIS, remote sensing, and environmental modeling
   (c) Partner with academic institutions to establish urban labs or observatories

Future Research Directions

While this book has made substantial contributions to understanding urban landscape resilience, several areas warrant further exploration:

- Longitudinal studies: monitoring landscape interventions over time to assess ecological, social, and economic impacts.
- Behavioral research: investigating how different demographics use and perceive green spaces, particularly under climate stress.
- Ecosystem services valuation: quantifying the monetary and non-monetary value of urban ecological systems to influence policy prioritization (Dong & Hauschild, 2017).
- Comparative south–south dialogues: expanding research to other cities in Asia, Africa, and Latin America that share governance and environmental parallels.
- Post-disaster urban landscapes: exploring how sustainable landscapes contribute to recovery in cities affected by floods, heatwaves, and socio-political unrest.

Conclusion

Cities today stand at a pivotal crossroads in their evolution. Faced with the compounding pressures of rapid urbanization, climate change, biodiversity loss, and social inequality, urban areas must choose between continuing down unsustainable paths of fragmented planning and ecological neglect or embracing a more integrated, equitable, and resilient future. At the heart of this decision lies the role of sustainable landscapes, which are increasingly recognized not merely as decorative or recreational features, but as essential infrastructure that underpins ecological health, urban livability, and climate adaptation.

Throughout this book, the case studies of Islamabad and Bucharest have demonstrated how the neglect of urban ecological systems can deepen vulnerabilities, especially among marginalized populations. Conversely, strategic interventions—whether in the form of reforestation, green corridors, community gardens, or small public urban green spaces (SPUGS)—can serve as catalysts for regeneration, both environmentally and socially. These examples reaffirm that landscapes must not be peripheral in planning processes, but rather central to the resilience agendas of contemporary cities.

Crucially, the transition to landscape-centric urbanism cannot be driven by planners and policymakers alone. It requires a collective commitment from a broad coalition of actors: citizens who advocate for and care for local green spaces; researchers who generate actionable knowledge; local governments that enforce equitable planning standards; and international bodies that align funding with sustainability outcomes. This multi-actor approach is key to ensuring that sustainable landscape planning moves beyond vision statements into tangible, place-based transformations.

Furthermore, the idea of landscape must expand to reflect social, cultural, and political realities. Urban green spaces are not just ecological zones—they are sites of memory, identity, negotiation, and inclusion. When planned inclusively, they can become platforms for dialogue, healing, and innovation. When ignored, they become symbols of inequality and institutional failure.

In conclusion, the future of resilient cities hinges on our ability to reimagine urban landscapes not as passive backdrops, but as dynamic systems that sustain urban life. This book calls for bold, interdisciplinary, and justice-oriented approaches to planning—approaches that respect the ecological complexity of cities and uphold the right of all citizens to live in healthy, equitable, and vibrant environments.

References

Agyeman, J., Bullard, R. D., & Evans, B. (2003). *Just sustainabilities: Development in an unequal world*. MIT Press.

Anwar, M. M., & Aziz, A. (2025). Mapping the land use dynamics and their impacts on sustainability: A geographical study of Islamabad, Pakistan. Journal of Urban Ecology and Sustainability (In press).

Dong, Y., & Hauschild, M. Z. (2017). Indicators for environmental sustainability. *Procedia CIRP, 61*, 697–702. https://doi.org/10.1016/j.procir.2016.11.213

Forman, R. T. T. (2014). *Urban ecology: Science of cities*. Cambridge University Press.

Gavrilidis, A. A., Popa, A. M., Onose, D. A., & Grădinaru, S. R. (2022). Planning small for winning big: Small urban green space distribution patterns in an expanding city. *Urban Forestry & Urban Greening, 78*, 127778. https://doi.org/10.1016/j.ufug.2022.127778

Kabisch, N., Korn, H., Stadler, J., & Bonn, A. (2017). *Nature-based solutions to climate change adaptation in urban areas: Linkages between science, policy and practice*. Springer.

Meerow, S., Newell, J. P., & Stults, M. (2016). Defining urban resilience: A review. *Landscape and Urban Planning, 147*, 38–49. https://doi.org/10.1016/j.landurbplan.2015.11.011

MIX
Papier aus verantwortungsvollen Quellen
Paper from responsible sources
FSC® C105338

If you have any concerns about our products,
you can contact us on
ProductSafety@springernature.com

In case Publisher is established outside the EU,
the EU authorized representative is:
Springer Nature Customer Service Center GmbH
Europaplatz 3, 69115 Heidelberg, Germany

Printed by Libri Plureos GmbH
in Hamburg, Germany